POPULATION, RESOURCES AND DEVELOPMENT

International Studies in Population

Volume 1

The International Union for the Scientific Study of Population (IUSSP)

The IUSSP is an international association of researchers, teachers, policy makers and others from diverse disciplines set up to advance the field of demography and promote scientific exchange. It also seeks to foster relations between persons engaged in the study of demography and population trends in all countries of the world and to disseminate knowledge about their determinants and consequences. The members of the IUSSP scientific group responsible for this book were chosen for their scientific expertise. This book was reviewed by a group other than the authors. While the IUSSP endeavors to assure the accuracy and objectivity of its scientific work, the conclusions and interpretations in IUSSP publications are those of the authors.

International Studies in Population (ISIP) is the outcome of an agreement concluded by IUSSP and Springer in 2004. The joint series covers the broad range of work carried out by IUSSP and includes material presented at seminars organized by the IUSSP. The scientific directions of the IUSSP are set by the IUSSP Council elected by the membership and composed of:

Population, Resources and Development

Riding the Age Waves – Volume 1

Edited by

SHRIPAD TULJAPURKAR

Stanford University,
California, U.S.A.

IAN POOL

Waikato University,
Hamilton, New Zealand

and

VIPAN PRACHUABMOH

Chulalongkorn University,
Bangkok, Thailand

Springer

A C.I.P. Catalogue record for this book is available from the Library of Congress.

ISBN-10 1-4020-3821-6 (PB) Springer, Dordrecht, Berlin, Heidelberg, New York
ISBN-13 978-1-4020-3821-6 (PB) Springer, Dordrecht, Berlin, Heidelberg, New York
ISBN-10 1-4020-3412-1 (HB) Springer, Dordrecht, Berlin, Heidelberg, New York
ISBN-10 1-4020-3464-4 (e-book) Springer, Dordrecht, Berlin, Heidelberg, New York
ISBN-13 978-1-4020-3412-1 (HB) Springer, Dordrecht, Berlin, Heidelberg, New York
ISBN-13 978-1-4020-3464-0 (e-book) Springer, Dordrecht, Berlin, Heidelberg, New York

---

Published by Springer,
P.O. Box 17, 3300 AA Dordrecht, The Netherlands.

Printed on acid-free paper

Printed in the Netherlands.

CONTENTS

PREFACE

SHRIPAD TULJAPURKAR
CYRUS CHU
ANNE GAUTHIER
NAOHIRO OGAWA
IAN POOL

Beginning in the last century and continuing into the 21st century, the populations of the world's nations have displayed large and long-lived changes in age structure. Many of these began with fertility change in the form of baby booms, busts or declines, and are amplified by declining mortality and by migration within and between nations. These age structural transitions have powerful effects on human affairs, from driving the demand for public and private goods and services for young and old, to determining the flow of resources across the different ages of the human life cycle. The consequences of age structural transitions vary in emphasis and detail, but not in significance, across the spectrum of nations in different stages of economic development. Demography will matter in this century not by force of numbers, but by the pressures of waves of age structural change.

In 1997 a committee of the International Union for the Scientific Study of Populations was charged with exploring age structural transitions and their policy implications. The committee brought together distinguished scientists to examine the key demographic, social, economic, and policy aspects of age structural change across a spectrum of nations at different stages of development. Readers will find a rich discussion of their work in "Riding the Age Waves", a series of three books, "Population, Resources and Development," "Allocating Public And Private Resources Across Generations," and "Responses to Aging in Advanced Industrial States." These volumes deal, roughly speaking, with developing economies, rapidly industrializing economies, and highly industrialized economies.

Developing countries face challenges ranging from building human capabilities and creating jobs to creating industry, infrastructure and institutions; all of these depend critically on the flow of people into key age groups of the life cycle (school, college, job seeking, marriage, parenthood, etc.). Because many developing countries are in relatively early stages of fertility decline, they are in the early stages of baby booms which will inevitably produce echoes – these manifest as age waves that last for two or more generations. These waves in turn create shifting flows of people into the key age groups, greatly complicating the task of managing the many components of development.

In the rapidly industrializing countries, population age structures typically reflect large fertility declines that occurred a generation or so ago, with resulting baby booms that have aged and are moving through the labor force. Fertility declines continue to have significant social and economic effects, e.g., on the growth of individual savings and thus aggregate capital investment, on declining family sizes and the shift from traditional family structures towards nuclear families, and on the thinning out of traditional kinship networks and changes in traditional forms of intergenerational support. Each of these has short and long run effects on government policy with respect to education, social services, welfare, old-age support, and macroeconomic and fiscal policies.

The highly industrialized countries confront the rapid decline of mortality leading to increases in the length of lives post retirement, the persistence of below-replacement fertility, and in many cases the long term age distortions of post World War II baby booms. In these countries, age structural changes pose challenges in terms of a long run decline in the labor force, a shifting balance towards high old-age dependency, and growing demands for old-age pensions and health care.

Several features distinguish these books from past writing on the subject: first, a joint examination of dimensions of age structural change that have often been considered in isolation from each other (for example, in developing economies, education, job creation, land use, health); second, the papers here bring together the many policy implications of these dimensions; third, the use of case studies to examine policy consequences and options of particular dimensions of change; fourth, the development of qualitative and formal methods to analyze the long term dynamic nature and consequences of age structural change.

The committee (Shripad Tuljapurkar, Cyrus Chu, Anne Gauthier, Naohiro Ogawa, Rafael Rofman, Ian Pool, Hassan Yousif) thanks the many people and agencies who made this work possible. In addition to the IUSSP, we thank the Asian Metacentre at the Singapore National University, the Academia Sinica in Taipei, and the Nihon University Population Research Institute in Tokyo. The editors of the individual volumes were Cyrus Chu, Anne Gauthier, Naohiro Ogawa, Ian Pool, Vipan Prachuabmoh, and Shripad Tuljapurkar.

ACKNOWLEDGEMENTS

This book has it genesis in a conference on age-structural transitions and their Policy Implications, held in Phuket, Thailand in November 8 to 10, 2000. That meeting was organized by an IUSSP (International Union for the Scientific Study of Population) Committee on Population Age Structure and Related Public Policy (Shripad Tuljapurkar, Chair, Cyrus Chu, Anne Gauthier, Naohiro Ogawa, Ian Pool, Rafael Rofman, Hassan Yousif) and supported by the Asian MetaCentre for Population and Sustainable Development Analysis at the National University of Singapore (Wolfgang Lutz, Verene Koh and Vipan Prachuabmoh represented the MetaCentre). An editorial committee selected a subset of the many excellent papers presented at the meeting, reviewed them, and requested revisions: the final revised papers are included here.

THE AUTHORS

John Bryant has a Ph.D. in Demography from the Australian National University. He has worked at Khon Kaen University, Thailand, the University of Otago, New Zealand, and the New Zealand Treasury. At the time his paper was written, he was undertaking a Wellcome Trust Post-Doctoral Fellowship in Population Studies. His research interests include demographic theory, health services, and population projections.

Angelique Chan is an Assistant Professor of Sociology at the National University of Singapore. She completed her doctoral degree at the University of California Los Angeles, and a postdoctoral fellowship at the University of Michigan, Ann Arbor. Her current research focuses on the social and economic consequences of ageing in Asia.

Napaporn Chayovan, Associate Professor, Deputy Director of the College of Population Studies, Chulalongkorn University. She studied sociology and demography at Chulalongkorn University, the University of London and the University of Michigan. In 1973 she received a Master's degree in Demography from the London School of Economics and Political Science, University of London, and a doctorate degree in sociology from the University of Michigan.

Socorro A. Gultiano is the Director of the Office of Population Studies of the University of San Carlos in Cebu City, Philippines, and is an Associate Professor of the Sociology–Anthropology Department of the University.

Gavin Jones is Professor of Demography at the Australian National University. Earlier, he worked for nine years with the Population Council and for many years as consultant in planning agencies and universities in Thailand, Malaysia, Indonesia and Sri Lanka.

Midea Kabamalan is a graduate student of sociology at the University of Hawaii. Her research interests include marriage, cohabitation, family and household formation, youth studies, and migration. She is currently involved with the 2002 Philippine Young Adult Fertility and Sexuality Study of the Demographic Research and Development Foundation and the University of the Philippines Population Institute.

Nan Li is a Research Scientist in the Program on Population, Policy, and Ageing at the Terry Sanford Institute of Public Policy, Duke University. Trained in mathematics and demography, his current research interests include stochastic mortality and population forecasts, modeling mortality at oldest-old ages, and measuring national-level disability risk. He received his Ph.D. from Xi'an Jiaotong University.

Wolfgang Lutz is Leader of the Population Project at IIASA (International Institute for Applied Systems Analysis, Austria), Principal Investigator of the Asian MetaCentre for Population and Sustainable Development Analysis (with headquarters in Singapore), Co-ordinator of the Global Science Panel on Population and Environment and Secretary General of the IUSSP (International Union for the Scientific Study of Population). His main interests are population forecasting, family demography and population–environment interactions. He recently published "The end of world population growth" in *Nature* (2 Aug 2001) and edited "Population and environment: methods of analysis", *Population and Development Review* Special Supplement 2002.

Ian Pool, Fellow of the Royal Society of New Zealand, is Professor of Demography, University of Waikato, Hamilton, Honorary Scientific Consultant (CICRED, Paris) and a member of a Panel on Scientific Capacities, International Council of Science, New Zealand. His research interests currently include age-structural transitions, population health, family demography and historical demography. He received his Ph.D. from the Australian National University.

Warren Sanderson, Professor of Economics, with a joint appointment in History, State University of New York at Stony Brook. His research fields are economic demography, labor economics, economic history. He received his Ph.D. at Stanford University in 1974.

Shripad Tuljapurkar, Professor of Biological Sciences and Morrison Professor of Population Studies, Stanford University. His research fields are mathematical and formal demography, economic demography, ecology and evolutionary biology. He received his Ph.D. at Portland State University in 1976.

Turgay Unalan is Associate Professor at Hacettepe University Institute of Population Studies, Ankara, Turkey. He is lecturing on International Migration, Internal Migration and Urbanization, and Data Processing and Social Statistics for Demographers. His current interest areas are migration, household demography, and social survey research.

Peter Urich is manager of the Planning, Governance and Sustainable Community Development Programme at the International Global Change Institute, University of Waikato. His research interests are in the areas of resource control and environmental change, community development, limestone environments and their conservation and organic agriculture. He received his Ph.D. at the Australian National University.

Peter Xenos is a specialist on Southeast Asia and publishes on both historical and contemporary topics. He recently published *Population and History: The Demographic Origins of the Modern Philippines*, and his current focus is the design of household surveys on young adult reproductive health concerns.

Introduction

CHAPTER 1. AGE-STRUCTURAL TRANSITIONS, POPULATION WAVES AND "POLITICAL ARITHMETICK"

IAN POOL
University of Waikato

VIPAN PRACHUABMOH
Chulalongkorn University

SHRIPAD TULJAPURKAR
Stanford University

This book is about age-structural transitions and their policy implications in the 21st century. The book's central thesis is that the population age structure in most countries around the world is undergoing cyclical, often irregular, change that will persist for many decades. These changing age structures will shape changes in human, social, institutional and economic needs and capacities, and pose significant challenges to policy makers. The description, analysis and prediction of these changes call for new methods and perspectives in several directions that are the subject of the chapters in this book. First, this book demonstrates that in virtually all countries over the coming decades age-structural transitions will affect cohorts at all stages of the life cycle, not just the young or the old. Second, this book shows explicitly how to use analyses of human capital and human development as a way of linking age-structural transitions to policy needs and constraints. Third, the book presents new methodological tools for the formal analysis and modelling of age-structural transitions, their demographic and socioeconomic effects, and their interaction with alternative policies. Fourth, the book uses new data to paint a detailed picture of particular age-structural transitions as they shape changes in demography, family, institutions, land use, and policy at levels of aggregation from individuals, to rural and urban segments within countries, to countries and regions.

From Ageing to Age-Structural Transitions

The ways in which demography informs policy analysis have roots going as far back as William Petty's remarkable essay on "political arithmetick" (Petty 1690). In the developed countries over the past two decades the looming spectre of population ageing has led to considerable work on the actuarial and fiscal effects of ageing on the economic well-being of nations (e.g., Lee and Skinner 1999; Lee and Tuljapurkar 1997, 2000). Such work on the fiscal effects of ageing embodies Petty's approach in modern form, and has an increasingly high public profile. Indeed, the emphasis on ageing suggests an orderly,

S. Tuljapurkar et al. (eds.), Population, Resources and Development, 3–10.

inexorable transition in population age structures: in response to long term declines in fertility and increasing life spans, all countries appear to be on a trajectory from "young" age pyramids with rapidly growing young cohorts to "old" age pyramids where growth is faster at old ages than young.

But demographic reality is more complex. Even in the rich Western countries, relative adult cohort sizes are still dominated by the fertility booms of the mid-twentieth century, while more recent fertility declines promise newer cohort imbalances in the future. In Asia and other rapidly developing regions, fertility declines came later than in the West but were often extraordinarily rapid, leading to large cohort imbalances across younger ages that will persist for many years. In Africa and other regions fertility rates have only recently begun to fall and large young cohorts will dominate the demographic landscape for some time. Mortality declines are widespread but are occurring at different speeds; and epidemics such as HIV/AIDs will take a considerable toll on young and working ages in many countries. Migration flows within and between countries add complexity to rates of cohort change. Overall, population age structures almost always display a complex and time-varying pattern – these are the "age waves" of the title of this book.

A recurrent theme in this book is that ongoing changes in fertility, mortality and migration have interacted to generate multiple population age waves within national populations, as discussed by Pool (Chapter 2, this volume). Cohorts all across the age spectrum, at critical life-cycle stages from school through work to retirement, display unequal rates of change and are likely to continue to do so for decades. An immediate consequence of this complexity is that a useful "political arithmetick" for the 21st century needs to consider many dimensions of demographic change and policy beyond those that are the focus of the literature on ageing *per se*.

Describing Age Waves and their Effects

In recognizing the significance of "waves" we follow an important trend in the literature whose focus has been mainly on the rich Western nations. In his path-breaking work, Easterlin (1987) showed that relative cohort sizes have a strong relationship to the fortunes of successive cohorts in the United States. Others have extended his work to examine the labour market effects and economic fortunes of cohorts as they track "waves" through time (for an example, see Macunovich 1999). Chesnais (1990) showed how age-structural transitions are reflected in uneven cohort sizes across a range of European countries; and Lindh and Malmberg (1999) relate these complex age-structural transitions to economic growth in the OECD countries.

The analysis of cohort imbalances has also been central to another important body of work on the economic development of the Asian "tigers" – the countries in East Asia such as Taiwan which experienced very rapid economic growth in the last decades of the last century. Higgins and Williamson (1997) and Lee, Mason and Miller (2002) show that in these countries fertility decline led to a substantial shift in savings behaviour, and thus to a possibly transient but nevertheless large rise in national savings. This in turn contributed to a rapid growth in capital and investment, fuelling and perhaps even making possible the striking rate of economic change in many countries. Lindh (1999) uses similar arguments to explore a range of linkages between economic growth and cyclical cohort dynamics.

These lines of work provide powerful evidence of the economic effects of age-structural transitions. But their emphasis is mainly on the economics of savings, capital formation and labour supply. What this book aims to do is broaden the perspective of "political arithmetick" to consider a much wider concept of individual and social welfare.

From Capital to a Human Development Perspective

Why do we need a broader perspective on age-structural transitions? The main reason is that uneven cohort sizes and unequal cohort growth rates translate into significant changes over time in the numbers of people at critical life-cycle stages. Life-cycle stages (and some corresponding policy variables) include the very young (primary education and infant/child health), the young (secondary and higher education and training), people at and shortly after entry into the labour force (job creation and spatial distribution), later stages in the labour market (job mobility, unemployment, housing, savings, health), retirement (pensions, private savings), and older people (private, familial and public support, health care, disability). Changes over time across cohorts are also accompanied by changes over space, as migration (rural–urban, international) acts to amplify or dampen age waves in different places. For example, ageing in the rural parts of the rich countries is far more advanced than in urban settings, and urban areas differ in their rate of ageing depending on the life-cycle structure of net migration.

Life-cycle stages are characterized by different levels of demographic "density" as discussed by Rindfuss (1991). The density here refers to life-cycle changes – biological, social, cultural – which individuals are subject to at any life-cycle stage, and related institutional and structural effects, behaviours, attitudes, norms, needs and demands.

Thus a life-cycle view focuses attention on important transitions in the development of human capital and important elements in the assessment of human welfare. When age-structural transitions lead to complex compositional change across the life cycle, policy must anticipate and accommodate the corresponding changes in human needs and capacities.

In this book we propose the use of a consistent focus on human development at ages across the entire spectrum of the life cycle. The goal is to evaluate human needs, capacities and welfare consistently across the entire age pyramid, and to incorporate the relevant compositional changes, especially age-structural and geographic. This approach identifies challenges and opportunities for many aspects of policy – social, economic, institutional, and 'population policy'.

New Perspectives on Managing Policy

The broad approach we discuss and exemplify has several advantages in identifying and analyzing policy changes. First, the shift away from central planning to sectoral approaches is enhanced by the need to formulate policies in response to age-structural transitions. Population waves, or troughs, entail cohorts of sizes that are relatively larger than historical, or smaller than historical, reaching key life-cycle stages. The needs of populations at these stages are normally met by a given sector, and, although the fit is not perfect, sectors tend to serve given age-groups (see Pool, Chapter 2, this volume). Current emphases on sectoral-specific planning favour the adoption of a life-cycle stage as an indicator of population-policy linkages. As Lindh and Malmberg (1999) have argued this permits better calibration of the interrelations between economic development and population change. Some past analyses showed little interaction between economic dynamics and cohort waves, partly because of the use of crude indices (e.g., total labour force growth), or a mismatch between the long time scale of gross population trends and much shorter economic planning horizons. In addition, analysis focused on the demand for services and human capital has meant that demographic factors were often subsumed into models relating to social sectors. But the more sophisticated methods we propose here will allow demographic factors to be taken into account more accurately in other sectors (e.g., taxation, welfare, transfers, and investments).

Second, as implied above, demographic factors have often been seen as driving the demand-side of the economy, but of limited importance for supply. Traditionally this was so even in the field of labour market economics, where the emphasis was on patterns of labour force participation, or by seeing labour force entrants as the substitutes of those retiring. But by viewing policy from an age-structural transition perspective, demographic factors can also be seen as being more central to the supply-side as well. Thus demographic dynamics delineate both challenges and opportunities, both burdens and capacities. In some countries today there is a 'window of opportunity' or 'demographic bonus' coming from inflated cohorts arriving at working ages, thereby reducing dependency burdens. But advantages of this sort can equally well translate into fiscal burdens if such waves are not managed well (e.g., if large cohorts reaching young working ages suffer disproportionately high levels of unemployment).

Third, the broad approach we espouse also directs attention to important co-variants of age-structural transitions. These include:

- social structural factors such as urbanization, economic modernization or shifts in patterns of social status;
- contextual factors such as changes in family structure and institutions;
- changing migration patterns and their relationship to labour deficits or surpluses.

Fourth, analysis of cohort dynamics draws attention to issues of resource availability and use. In this volume Urich and Gultiano (Chapter 11) consider the relationship between demographic change, life cycles and land use, while Chan (Chapter 12) relates cohort changes to housing supply and land use. Another example of a broader interaction between population structure and resources is provided by Liu et al. (2003) who examine the effects of household dynamics on resource consumption and biodiversity.

Comparative Analysis of Waves

The first section of this book has chapters by Pool, Jones, and Xenos and Kabamalam which provide comparative descriptions of age waves and their implications from distinct perspectives. Pool presents a wide ranging review of cohort dynamics, focusing on rates of change between cohorts and on comparisons across rich and developing countries. He shows that some cohort flows are properly seen as echoes of a past fertility transition towards replacement levels, whereas others are irregular and reflect multiple shifts in fertility and mortality over time. He highlights the significance of young cohorts around the world, and their contribution to population momentum – he argues that momentum is not merely a matter of contribution to long term growth but should be measured in terms of cohort-specific contributions to shorter term population change across life-cycle stages.

Jones examines the nature and consequences of the echo effect that persists for generations after fertility declines. He discusses the economic performance of Asian countries in terms of human capital investments. Much of his paper explores the effects of age-structural change on the challenges of increasing educational attainment and matching cohort flows to job creation over time. He shows that differences in demographic and institutional factors lead to different policy challenges.

Xenos and Kabamalam expand on Pool's general point about the importance of youth with a rich comparative analysis of cohort dynamics across Asian countries. They start with an instructive review of long term changes in young cohorts across history, and the social, behavioural and economic characteristics of young cohorts. They provide a demographically detailed look at the current and projected status of youth, using an informative array of demographic indices.

Methodology and Theory for Age Waves and Policy

The second section of this book sets out new directions in the analysis of populations that are undergoing age-structural transitions, and in relating these transitions to policy questions. The papers presented strike out in different directions in what is a new area. Li and Tuljapurkar present a general theory for the analysis of population waves that result from a transition in fertility and/or mortality rates. This work provides a comprehensive extension of stable population theory to cases in which populations complete a transition from high to replacement fertility. They illustrate their methods by examples. A significant aspect of these methods is that one can analyze population momentum in realistic situations that go well beyond the classical methods developed by Keyfitz.

John Bryant considers a problem that is of great significance in many countries that are either in or have just completed a fertility transition: how can one estimate the potential level of familial support for the elderly? He points out that complex methods of doing such an analysis do exist, but require very detailed data and models. Instead he presents a simple, elegant and powerful solution which uses macro-level data. He illustrates this in the case of Thailand by computing the projected distribution of elderly women by the numbers of living children 25 years into the future. This type of analysis is a way of getting at an "individual" dependency ratio, and is important in the analysis of old age support and policy.

Lutz and Sanderson develop two interesting theoretical tools, one that extends demographic analysis and one for policy analysis. They start by defining, analyzing and illustrating the use of cohort-specific growth rates. This is precisely the sort of quantitative measure argued for by Pool in his comparative chapter, and provides a novel way of describing life-cycle dynamics across the age structure over time. Next they consider human capital investment in terms of a model of educational attainment and the relationship between income and education, and show how policy interacts with demographic change to affect human welfare over time. Their analysis provides a framework for thinking about long-term sustainability in terms of human welfare.

Finally, Tuljapurkar shows how one can use the dynamic analysis of Li and Tuljapurkar to examine the demand side of changing age waves. He considers in particular changes in education, health care, and pension expenditures over time in a population that experiences a transition to replacement fertility. This type of analysis reveals the quantitative shifts in expenditure and thus the fiscal demands that age waves will impose over time. He discusses ways in which these dynamics can be made endogenous to incorporate feedbacks between demographic and socioeconomic change.

Country-Specific Analyses

The last four chapters turn to substantive analyses of particular countries. These chapters are especially valuable because they illustrate the many dimensions of policy that are influenced by age waves. Chayovan considers the transition to an ageing population structure in Thailand, using data from two national surveys. She reports on many dimensions of the welfare of the elderly, from labour force participation to economic wellbeing and sources of old age support. Her data also reveal sex differences and urban–rural differences. She discusses the policy challenges that the government will likely face in terms of the welfare of the elderly, and the competing demands on governmental resources.

Unalan provides a rich picture of the demographic transition in Turkey, which underwent a series of transitions that lowered fertility rates significantly by the 1990s. In keeping with this book's theme, he goes beyond a description of the demographic transition to present a detailed description of changes in household and family characteristics. He documents a shift towards nuclear families and also in the age structure of households. These institutional and social factors have significant implications for policy with regard to housing, education, and transfers.

Urich and Gultiano examine the relationship between age waves, urban–rural migration, and agricultural land use in the Philippines. They use survey data to describe the dynamics and movement of youth, and relate these to the level and intensity of land use in agriculture. They discuss the factors that shape educational attainment and migration, the flows of income between parents and children, and policy challenges in education, agrarian reform, and migrant welfare.

Finally, Chan considers ageing in Singapore, using both forecast and longitudinal data. She uses these data to dissect the needs of the elderly by age, health, and wealth, providing a careful and nuanced analysis of the needs and capacities of the ageing population. She discusses the relationship between the changing structure and role of the family, and the determinants of welfare for the elderly.

Future Challenges

This book is a beginning to the life-cycle analysis of changing demographic structures and their effects on policy. Much needs to be done in this area as demographers and other social scientists confront the reality of age waves in the coming decades.

Conceptual and theoretical work is needed in describing and projecting age waves and their momentum effects. The papers in this book make a start but many open questions remain, including:

- how to analyze irregular oscillations and disordered cohort effects and the distances between waves;
- how to measure the flow-on effects of waves and disordered cohort flows, to quantify institutional as well as population-wide transitions that interact with age structure;
- how to compute momentum effects from cohort dynamics occurring simultaneously in several generations;
- how to describe and analyze welfare (and utility) functions that weight dynamically changing cohorts across the life cycle;
- how should one analyze convergence or divergence in age-structural transitions;
- is the convergence in demographic structures in many forecasts reasonable, or do we need better assumptions (Demeny 1997) and stochastic forecasts (Lee and Tuljapurkar 2000)?

The substantive analysis of age-structural transitions raises many questions that call for more analysis including:

- the relationship between cohort-specific needs and capacities and the sustainability of a welfare state;
- the effect of changing age structures on natural resources and ecosystems;
- the connection between age waves and social change in families, households, and at higher levels of aggregation;
- the effects of immigration flows within and between countries across life-cycle stages and cohorts.

Some of these topics have already generated interest in particular settings or particular countries, but our analysis of age waves suggests that they will be keys to the ways in which demography can help us to understand and respond to the future.

References

Chesnais, J-C. (1990), "Demographic transition patterns and their impact on the age structure." *Population and Development Review*, 16:327–336.

Demeny, P. (1997), "Replacement-level fertility: the implausible endpoint of the demographic transition." In: *The Continuing Demographic Transition*, edited by G.W. Jones et al. pp. 94–110. Clarendon Press: Oxford, England.

Easterlin, R.A. (1987), *Birth and Fortune: The Impact of Numbers on Personal Welfare*, 2nd edition. 235 p.

Higgins, M. and Williamson, J.G. (1997), "Age structure dynamics in Asia and dependence on foreign capital." *Population and Development Review*, 23, 2:261–93.

Lee, R., Mason, A., and Miller, T. (2002), "Saving, wealth, and the transition from transfers to individual responsibility: the cases of Taiwan and the United States." *The Swedish Journal of Economics*, 105:339–357.

Lee, R. and Skinner, J. (1999), "Will ageing baby boomers bust the federal budget?" *Journal of Economic Perspectives*, 13:117–140.

Lee, R. and Tuljapurkar, S. (1997), "Death and taxes: longer life, consumption, and social security." *Demography*, 34:67–81.

Lee, R. and Tuljapurkar, S. (2000), "Population forecasting for fiscal planning: issues and innovations." In: *Demography and Fiscal Policy*, A. Auerbach and R. Lee, eds. Cambridge University Press: Cambridge.

Lindh, T. (1999), "Age structure and economic policy: the case of saving and growth." *Population Research and Policy Review*, 18:261–277.

Lindh, T., Malmberg, B. (1999). "Age structure effects and growth in the OECD, 1950–1990." *Journal of Population Economics*, 12:431–449.

Liu, J., Daily, G., Luck, G., and Ehrlich, P. (2003), "Effects of household dynamics on resource consumption and biodiversity." *Nature*, 421:530–533.

Macunovich, D.J. (1999), "The role of relative cohort size and relative income in the demographic transition." *Population and Environment*, 21:155–192.

Petty, W. (1690), *Political arithmetick, or a Discourse Concerning the Value of Lands, People, etc., with a Dedicatory Epistle*, by C. Petty and B. Shelburne. Available on the web at http://socserv.mcmaster.ca/econ

Rogers, A. and Woodward, J. (1988), "Multiregional projections of the U.S. population: 1980–2080." *Population Program Working Paper*, No. WP-88-5. 56 pp. University of Colorado, Institute of Behavioral Science, Population Program: Boulder, CO.

Rindfuss, R.R. (1991), "The young adult years: diversity, structural change, and fertility." *Demography*, 28:493–512.

Part I

Issues and Patterns

CHAPTER 2. AGE-STRUCTURAL TRANSITIONS AND POLICY: FRAMEWORKS

IAN POOL
Population Studies Centre, University of Waikato, New Zealand

1. Dynamics and Structures

Much of the focus in demography over the last few decades has been on population dynamics, with far less attention given to structures, particularly age-structures. The recent growth of interest in ageing has, however, highlighted two key issues:

- that this particular area of demography has been relatively neglected; and
- that, at least for the next few decades and for the world outside Europe and Japan, ageing is not the only, or even the most important, form of age-structural change.

Age-structural changes occurring in different countries follow broadly similar patterns. They also seem to be evolving in a generally similar direction towards ageing over the long term, albeit at different speeds and from different starting points. All this is not surprising as these changes are the product of demographic transitions, which also follow similar general paths. It is appropriate therefore to view age-structural changes as 'transitions' and to attempt to document the most common patterns. This will permit us to develop a model that could be added to the existing family of 'transition' models in related fields, as well as to the classical demographic transition model (Notestein 1945). This is one of two central concerns in this paper.

In this context there is one conceptual problem that I would like to draw attention to. The term 'ageing' is an ambiguous one, often confounding the whole process of age-structural transition that reaches its final phase with structural ageing, with the restructuration attendant on reaching high proportions at older ages. It is for this reason that I have referred to the former as age-structural transition and the latter as ageing *per se*.

Age transition and other related models are of more than purely scientific interest. These sorts of models can form the underlying assumptions for forecasts in applied development areas. In spite of this potential, much of policy analysis and formulation takes little account of age-structural changes and of differences between populations. Yet the age-structure/policy inter-relationship may well be the strongest and most direct link between population change, policy formulation and development strategies. This point will be developed later: the policy implications of age-structural transitions constitute the second major concern of this paper.

S. Tuljapurkar et al. (eds.), Population, Resources and Development, 13–39.

2. Issues Underlying the Analysis of Age-Structural Transitions

In order to analyse and document age-structural transitions, and to assess their potential impacts on policy, there is a need to formulate frameworks and models at three levels: theoretical, conceptual and analytical. This paper will focus on conceptualisation.

The major contributions to the theoretical elaboration of age-structural effects have come from mathematical demography (e.g. Keyfitz 1971; Coale 1972) – a domain that, arguably, has taken our understanding of this aspect of population change further than the more empirical branches of the discipline. This is because much of the empirical work has involved merely cross-sectional analyses that employ simple percentage based indices. The use of cohort-component analysis as a baseline for projections has been rather more sophisticated; but here the focus has been on generational trends in fertility, mortality and migration, rather than cohort age-structural trends: structure has been an outcome rather than a central concern.

One can contrast this with the attention given since the 1960s to both the theoretical underpinnings and the methodological tools for the indirect estimation of vital rates. As a consequence, our empirical knowledge base on population size and dynamics across the globe has been advanced enormously. Ironically, however, underlying and sustaining these developments has been work on stable population theory, and other related aspects of age structure. My paper does not attempt to build on this theoretical work, but I do wish to signal an urgent need to return to this area in order to meet emerging needs, not over ageing *per se* but the other phases of age-structural transitions to be outlined below.

There is a parallel need, but it is also one that is beyond the scope of this paper – to develop more refined and rigorous conventional methodologies than those existing at present, as documented in standard texts in the field (e.g. Shryock and Siegel 1976). The issue of cohort oscillations would seem to be an area of immediate critical importance.

While this paper does not attempt to develop further the existing methodologies, it is worth noting, however, something that is self-evident yet important for the rest of my presentation. In demography we conceptualize trends in terms of period changes, so that our measurement systems have been most highly developed in this area, based on total population sizes and vital and migration rates, or related measures, calculated for events occurring over a particular time period. When we turn to age-structural analysis, however, the unit of reference becomes the age group, and more usefully the birth-cohort, and its changes in size over its life span due to mortality and migration. It is the absolute or relative sizes of generations that become central to any analyses.

When age structure is analysed its transitions are typically seen as almost independent of, and the outcomes of, classical 'demographic transitions'; they are less frequently seen as proximate determinants of population dynamics and of demographic transitions, even though this does occur. In fact, natural increase and age-structural changes are intimately interlinked "because growth rates are generated by the interaction of fertility and mortality patterns with the age structure . . . " (Frejka 1982: 450).

The question thus arises: should age-structural transitions be conceptualized as exogenous to demographic transition frameworks? Or should they instead be central to, and not merely implied by, demographic transition models?

In its singular interest in dynamics alone, the classical demographic transition model differs from other transition models, such as the epidemiologic (Omran 1982), mobility (Zelinsky 1971), nutritional (Drewnoski and Popkin 1997), and industrial labour force transformation (Chenery and Syrquin 1975; Madison 1982; see also Loriaux 1990: Figure 6 "Convergences des futurs: Quatre Schémas théoriques pour quatre mondes liés", especially technologies and employment). In contrast, these other models also cover structural changes in relation to shifts in dynamics (this issue is discussed further in Pool 2000). The age-structural transitions should, in fact, be seen as constituents of comprehensive demographic transition frameworks, sitting alongside and integrated with what should be more precisely termed the 'natural increase transition', the two sub-components of which are fertility and mortality transitions.

Thus as elaborated at present, the demographic transition model encompasses only part of the experience of societies, even for population trends as very narrowly defined. This means, incidentally, that substantive analyses and debates over transition theory are not in accord with theoretical elaborations of demographic change (Keyfitz 1968: especially Chaps. 7 and 14; Coale 1972: Chaps. 4–6). To formulate a more realistic model, grounded in demographic theory, it would be necessary to redesign the transition model as in the schema below.

DEMOGRAPHIC TRANSITION

↓ ↓

Age-structural transition **Natural increase transition**

↓ ↓

Fertility transition **Mortality transition**

This raises yet another question. There are now concerns about the decreasing utility of the classical model, even in its conventional form, for populations whose levels of reproduction now fall below replacement (see, for example, Jones and Douglas 1997; Demeny 1997). This question is also outside the scope of the present paper, but for the moment I will assume that the last stage of the classical model need not terminate when the population reaches exact replacement. Obviously, the implications of this are an issue of major theoretical and policy interest, as they raise spectres such as the rectangularisation of the age pyramid (Lévy 1998), and eventually inverted pyramid shapes. The concern of this paper is what happens as a population moves towards this stage, rather than what occurs at the end of the age-structural transition.

The natural-increase transition will go in advance of the age-structural transition "as the latter cannot be modified quickly" (Frejka 1982). Thus a significant duration could occur between the reaching of deep troughs of sub-replacement fertility, as is occurring in Italy or Spain today, and the occurrence of inverted age pyramid structures. It is a truism, but worth noting, that at that stage the capacity of the population to reproduce would be impaired by its age structure, even if a return to high levels of age-specific fertility were to be achieved; and crude death rates would be very high even if age-specific risks were low.

3. Mapping Age-Structural Transitions

3.1. INTRODUCTION

It is now necessary to ask what an 'age-transition framework' might look like, in which quest I will follow convention by modelling the transition as a series of phases and sub-phases. What I present here should be seen as a first attempt at this exercise. It should also be seen as a call for more theoretical work to update the classical studies of greats such as Keyfitz (1971) or Coale (1972). Such initiatives would also be highly applied as they would provide a lead for the development of more powerful methodologies.

Younger generations of demographers, among them some of the more distinguished in the field, have made significant contributions to the literature in this area. For example, work initiated with Coale (Preston and Coale 1982) was extended by Preston (1986), provoking a vigorous debate with Wachter (1988) and a response by Preston (1988). More recently, Horiuchi (1995: 59–60) has provided evidence about the way that events "that occurred several decades previously can exert substantial impacts on the current growth rate".

Perhaps most central to the current paper is work by Chesnais. He has documented the demographic transition and has shown how differential velocities of change have had markedly different effects on broad age-structures. His initial aim was to explain how "[t]he demographic transition is the process of modernization of the reproductive behaviour in human populations. In the typical case, the first stage of this process is mortality decline, the second is fertility control" (Chesnais 1990: 327). Fertility changes pass through two stages, "limitation of marriages, then limitation of births" (Chesnais 1986: 1061). He formulated a measure of the effects of these different determinants of replacement by what he termed "a synthetic index [the population multiplier] that characterizes the pattern of demographic transition.... This index is the number by which the population is multiplied during the transition between the pre-transitional phase (high mortality, high fertility) and the post-transitional phase (low mortality, low fertility)." (Chesnais 1990: 327).

Chesnais recognised that during a transition "the increase in population is not evenly distributed by age" (1990: 332). Thus he constructed "age-specific transitional multipliers" for broad age-groups for populations undergoing different transitions, and compared those for the young (0–14 yrs.) with those for persons aged 65+. The faster the transition, the higher the age-specific multipliers. For example, the schema below compares those for France and Kenya.

Transitional multipliers for:	**France**	**Kenya**
Total population	2	20
Ages 0–14	1.5	10
Ages 65+	10	200

Source: Chesnais 1990: 336.

This present paper extends Chesnais' work in two ways. Methodologically, it uses more refined age-groups. More importantly, it views the demographic transition, at least changes as producing disorderly and even turbulent trends (a point Chesnais had already presaged), by noting that at least in the pre- and post-transitional phases, "the notion of equilibrium . . . " should be seen only as relative ("La notion d'equilibre . . . merite d'être fortement relativisée"). Of course, he was more concerned with conditions of growth *per se*, rather than age-structures, but clearly diverse patterns of growth would also imply variations in structural transitions (1986: 1067).

Seminal as these studies might be, the impression remains that they arise spasmodically and have been marginal to the central interests in the field. To the extent that research agendas are driven by applied funding agencies, this work has appeared outside the major foci emanating from fora such as the Cairo conference, which emphasized reproductive choice and the empowerment of women. Yet this area in which much of the discipline has merely dabbled may prove to be the most fruitful, intellectually, and highly applied in the near future.

There are major questions that new theoretical initiatives might address, but which are beyond the scope of this paper. Suffice it to say that age-structural transition is also a problem that has methodological, substantive and policy implications. As I will show below, for the next few decades most of the world's age structures will be subject to "disordered cohort flows" (a term coined by Keyfitz). This is an experience that, in the provocative words of the Belgian demographer Loriaux, poses a challenge to our discipline. This is because demographers "have a sort of instinctive attraction for order, stability and equilibrium . . . [that] frequently causes us to lose sight of the fact that the evolution [of populations] is primarily an 'affair' involving disorder, instability and dis-equilibrium" (1990: 10; translated by the present author). Equally well, much of the literature on ageing, especially in applied fields, carries with it more or less the notion of a monotonic evolution towards ageing. The reality may instead be one of the effects of turbulence coming from significant cohort oscillations, co-varying in time with a structure also inexorably marching towards ageing.

Dittgen (2000: 24–25) has added another twist to this notion of turbulence for those populations whose passage to ageing *per se* would seem the most gentle and assured. Yet, so he has argued, the developed countries of Northwestern Europe have age pyramids that "are very irregular particularly where natality has dropped radically. These irregularities are costly financially and socially. . . ". When cohort sizes produce trough (vs. wave) phases, immigration will be necessary to meet gaps "not for demographic imperatives, but for economic ones that will fluctuate . . . . These [migratory] movements will only serve to accentuate the chaotic trends in the pyramids" (translation present author).

There are further dimensions to this. Firstly, the impacts on policy come from the fact that 'instability' accompanying 'disordered cohort flows' will characterise different phases of the life cycle at times as cohorts of varying sizes flow through these phases. But, as the American demographer Rindfuss has shown (1991), the life-cycle stages are not uniform in terms of what he calls 'demographic density' – the different events that will be experienced at each stage. He singles out the youth age group as one at which there is a high density of events occurring over a relatively short age-span. These include:

- finishing formal education and starting a job;
- passing through puberty and reaching physical maturity;
- geographic mobility – the 'floating populations' are typical at these ages;
- the commencement of relationships, then marriage and then the starting of a family.

Secondly, however, any one population may simultaneously experience the effects of disordered flows occurring at different stages of the life cycle. Beyond this, various regions within countries will have different age-structural trajectories, as has been demonstrated for New Zealand (Lepina and Pool 2001). These points have important implications not only for theory and methodology in this area but also for policy, as will be discussed in the last section of my paper.

For the moment, however, I wish to set out an age-structural transition model that draws heavily on, but elaborates a paper I presented earlier. I will turn frequently to the metaphor of waves, 'population waves' as Keyfitz called them (cited also in Rowland 1996). The metaphor 'ola' or Mexican wave is also appropriate, particularly as it is Mexico that is almost a prototypical example of this aspect of transition (Pool 2000).

3.2. AN AGE-STRUCTURAL TRANSITION FRAMEWORK

The framework presented here (drawn from Pool, 2000) posits a transition of three phases and a number of sub-phases:

1. The initial phase: ***Phase of Simple Momentum***, is divided into three sub-phases:
   - Stability and quasi-stability including stationarity
   - Gradually changing or consistent momentum
   - Rapidly accelerating momentum
2. The intermediate phase: ***Phase of Population Waves***, is divided into three sub-phases:
   - Simple oscillation and decelerating momentum
   - Double oscillations } A phase of disordered cohort flows
   - Multiple oscillations } A phase of disordered cohort flows
3. The final phase: ***Ageing***, is divided into three sub-phases:
   - Total population growth from momentum effects
   - Stationarity
   - Decrease in population size

Momentum is a key term in this framework. It has been defined rather narrowly in some demographic analyses as "a property whereby populations change their growth rates in a relatively smooth fashion" (Frejka 1982). Here I have preferred to use three closely linked, broader and somewhat more utilitarian definitions derived from physics. At their core is an element imputing the notion of impetus (*New Shorter OED*, 1993: vol. A–M, 1810):

> The quantity of motion of a moving body, equal to the product of the mass and velocity, of the body;... The effect of inertia in the continuance of motion; *impetus* gained from movement; *fig*. strength of continuity gained from an initial effort.[1]

---

[1] The seminar at which an earlier version of this paper was presented, Phuket, 8–10 November 2000, by consensus adopted the OED definition, especially the notion of *impetus*.

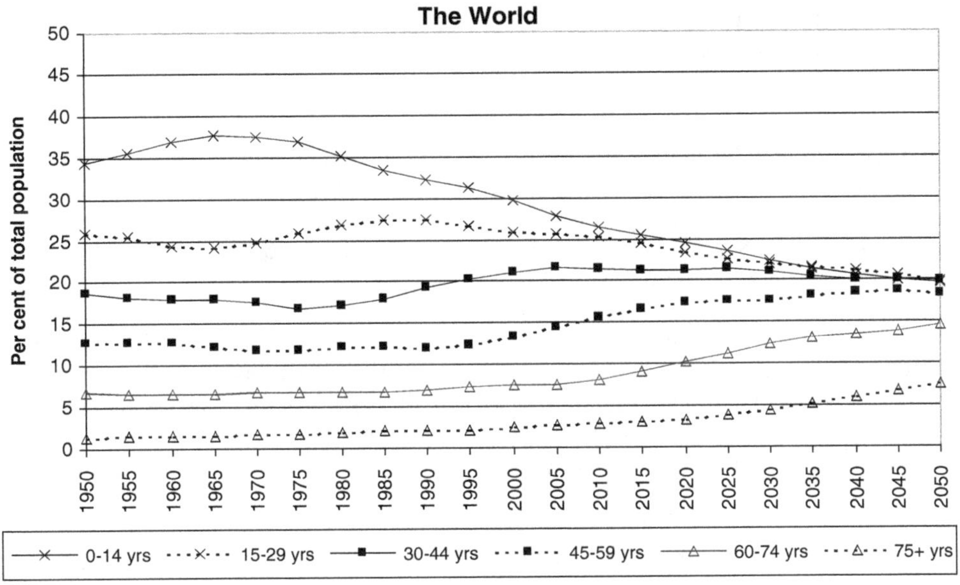

Figure 1. Percentage age distribution, functional age groups.

In Figure 1 data are presented on the world as a whole. It represents Phase II, and also demonstrates some modest waves and ebbs. Later I will use case study countries that can be compared for levels, trends and patterns with the world as a whole.

Basic demographic data on the case study populations drawn on this are given in Table 1. This table and all other data presented here come from United Nations projections, and thus are dependent on the assumptions on which these estimates are based, notably over

Table 1. Case study countries: basic demographic data.

	Total fertility rate (TFR)		% 65+		Av. annual rate of growth (%)	
	1970–75	1990–95	1990	2020	1970–75	1990–95
Congo Democratic Republic	6.3	6.7	3	3	2.7	3.8
Mexico	6.5	3.1	4	8	3.1	1.8
Thailand	5.0	2.1	4	9	2.6	1.4
China	4.9	1.9	6	12	2.2	1.1
India	5.4	3.7	4	7	2.2	1.9
USA	2.0	2.1	12	16	1.0	1.1
Russian Federation	2.0	1.5	10	16	0.6	−0.02
France	2.3	1.7	13		0.8	0.5

Source: UN (2000).

fertility declines. Thus, the results presented below are not in any sense definitive but are intended to raise the relevant questions.

Initial Phase

Some societies may be subject almost to a *pre-phase* before the initial phase of declines in numbers or what can be called negative momentum. This typically occurs for isolated populations that are immunologically virgin coming into culture contact with other groups who introduce diseases that were previously absent (Crosby 1986; for example, the Maori in New Zealand, Pool 1991: Chaps. 3–5).

The first sub-phase is one in which no changes occur in rates of growth and corresponds to situations also of stability or quasi-stability. While the second involves change it could also be known as one of quasi-stability, as these will be gradual and consistent. The third part of the initial phase would be when growth accelerates rapidly, normally as a function of rapid declines in mortality especially during infancy and childhood, while fertility rates remain high.

The third sub-phase requires particular attention, for it is the stage seen since World War II for societies starting their natural increase (see Dyson & Murphy 1985) and age-structural transitions. It is still the phase which best characterises much of Africa, although today parts of its southern and eastern regions and its small-island states are going through the fertility declines which would (or have) move(d) them into the intermediate phase. In this paper this sub-phase is represented by the Democratic Republic of the Congo (see Figure 2, panel A). Beyond this, a new factor has emerged: the effects on structures of mortality coming from the HIV/AIDS epidemic which is highly age-specific in its incidence. This may be the first instance in recorded demographic history, in normal periods (outside war) where survivorship levels at the young adult and middle ages have declined so radically and on such a wide scale. This phenomenon is not built into the framework presented here, but must be seen as an issue of utmost importance.

This sub-phase resembles more a tidal wave (Pool 2000; see also Loriaux 1990) than a normal wave. Its effects will be felt for a very long time as the population at each life cycle stage more than doubles over a short period (say 16–23 years). At present it is affecting the provision of education and entry into the labour market. This is also the situation over the longer term even for basic commodities such as food (Collomb *et al.* 1996). The United Nations projections (1992, 1998 revisions) show that at all ages, and across broad age groups, the poorest countries will see more than a fivefold increase in population size between 1950 and 2050. For the industrialised countries at all ages this will be only an increase of 1.7 times. For the age group 60–69 years, the very population around whom much of the concern over increased ageing is directed, the increase will still only be by 2.9 times.

Intermediate Phase

The situation just described could have been applied to describe much of the world until the 1970s or 1980s. But from about 1980 this changed, as fertility rates had declined rapidly across many Asian and Latin American countries. In the decade 1950–60 the age group

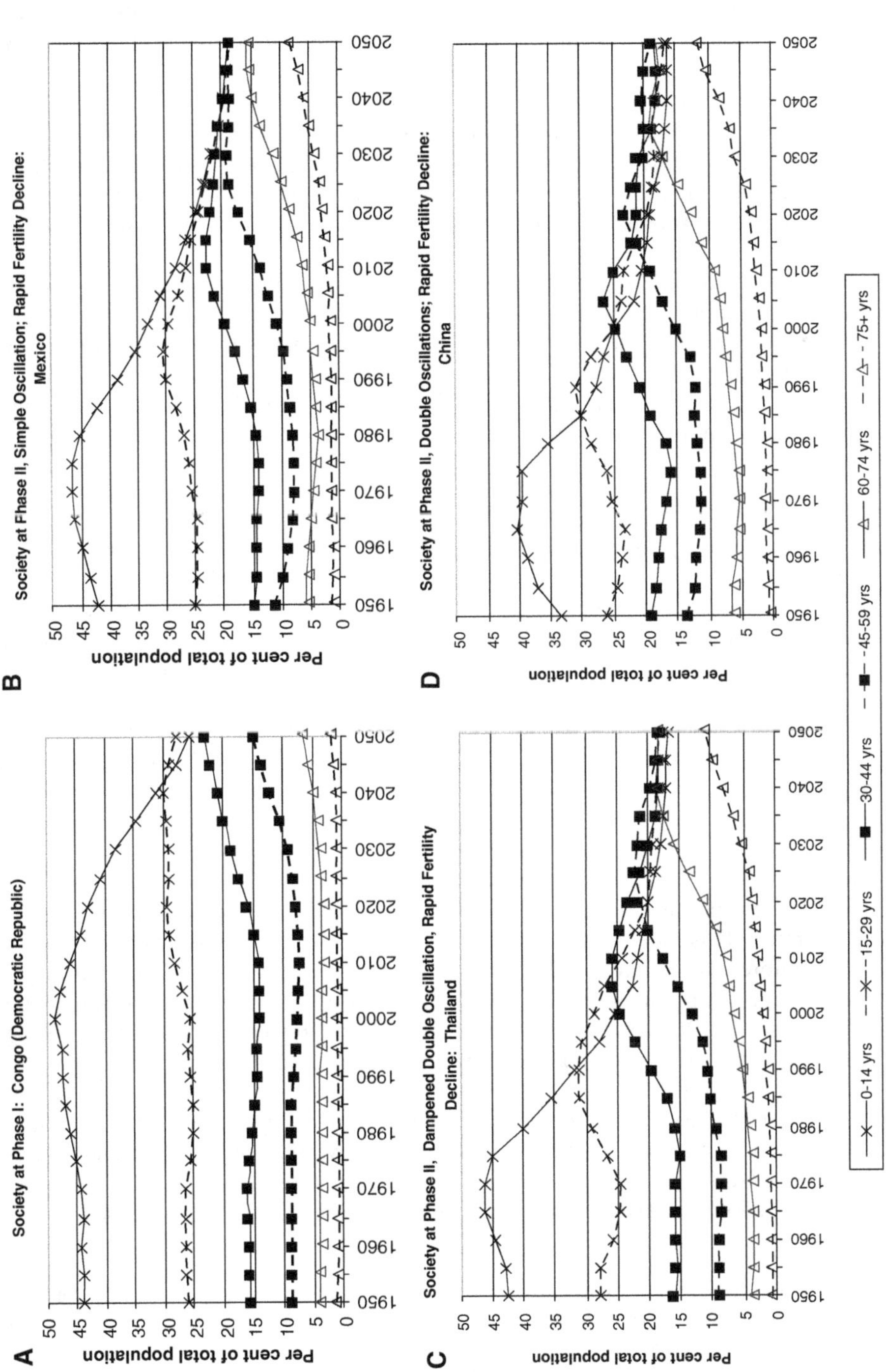

Figure 2. Percentage age distribution, functional age groups in Congo, Mexico, Thailand and China.

0–9 years, corresponding roughly to birth cohort size, had grown by 194 million children worldwide; from 1960–70 by 178 million, but over the decade 1970–80 by only 84 million.

By the 1990s only a small proportion of the global population, mostly in Africa, lived in countries at an initial phase of the age-structural transition. In the same period, a few European countries and Japan were entering the last phase of the age-structural transition. By contrast, most nations of Asia, most of the Americas, including Canada and the United States, and Australasia were at the intermediate phase, and that situation persists today.

Effectively, between the 1970s and the 1990s the world went through a massive shift between two demographic regimes:

- The old regime whose population trends were driven primarily by the natural increase transition, producing high rates of growth;
- A new regime which saw a rapid deceleration in these rates, but producing a series of age-structural mutations (Loriaux 1990) never experienced in this way in history before.[2]

The net result is that increasingly the demographic changes in most populations of the world are coming from compositional effects rather than natural increase effects.

The change in regimes came about because there was a shift across much of the world, especially among the more developed third world countries, from rapidly accelerating to rapidly decelerating momentum, setting in motion a trough following the wave created at the end of the initial phase. In the developed countries, the rupture between the two regimes was far less severe, yet was sufficiently marked for the European demographers to talk about a 'Second Demographic Transition' (Van de Kaa, 1987, 1988). This is a restructuration by age, of which ageing *per se* is merely the last phase. But before that situation is reached there will be fluctuations, often in the form of a series of waves, at younger ages.

Firstly, there is a sub-phase of simple oscillation, due to a shift from accelerating to decelerating momentum. This 'ola', or Mexican wave, which has its genesis in the period of very high fertility before rates began to fall over much of the world, will have/is having major impacts on all aspects of human life. For the world as a whole it commenced its impact on ages 0–14 years in the 1960s (even before), then moved on to hit ages 15–24 years, shifting today to have its main life-cycle impact on intermediate ages (25–44, or 45–64).

This 'ola' has a secondary impact: the inflated generations will themselves produce large birth cohorts even if total fertility rates are relatively low and declining gradually as is shown in Table 2 (say two to three births per couple). This trend itself results in further momentum, and another wave, often termed an 'echo effect'. In turn the echo cohorts

[2] Migration could, of course, have this effect under extreme circumstances. One of the two most rapid inflows of Europeans to New Zealand (90,000+) occurred in the early 1860s when the population base was just over 120,000. The inflow of single males and also of young couples was to distort the structure for decades after this. Another example is the effect of the 'Great Leap Forward' in Yan Hao's paper on China at the conference in Phuket, Thailand.

Table 2. Absolute growth (millions) in the age group 0–9 years; child–woman ratio (Population 0–9/women aged 15–49), world.

Year	TFR	Growth (mills.) in population 0–9 yrs.	Child–woman ratio
1950	5.00	...	0.98
1960	4.98	194	1.14
1970	4.46	178	1.15
1980	3.64	84	1.00
1990	3.26	128	0.90
2000	2.90	134	0.85
2010	2.62	42	0.76

Source: United Nations Projections, 1992 Revision.

could produce an echo to an echo, although if fertility is reducing their size will also be decreasing. Together, however, these echoes give rise to double- or multiple oscillations, spaced roughly a generation apart. These effects are shown in Table 2. It takes the total fertility rate (TFR) for the world and compares it with a simple statistic, the growth across the world of the population aged 0–9 years. The age-structural impacts of change in the child population are shown in Table 2 by data on the child–woman ratio. This shows that over the last 50 years fertility levels have dropped, slowly, then quickly and then more slowly again; yet momentum effects have continued, and have had a flow-on effect into intergenerational dependency burdens. Moreover, fluctuations in the absolute growth of the child population can occur when fertility rates are declining.

The decline in the fertility rate is one of the success stories of past population programmes (as narrowly defined at the Cairo Conference in 1994, see UN, 1995). What is surprising is that the Cairo document paid scant attention to the consequences of this success story: the severe structural changes resulting from population waves. Cairo chose instead to take up once again what had become a somewhat *passé* battle against 'galloping growth'. The point to which I will return later in Part 5 of this paper is that the effects of 'olas' could threaten the implementation of programmes that attempt to achieve a better balance between population growth and development.

Moreover the effects of compositional changes are not spread evenly across regions. For example, the growth of the child population contributed almost nothing to the overall increase in the size of the population between 1955 and 1990 for developed countries, much of the increment (62 percent) coming from momentum effects at ages 25–64. Over the period 1990–2025, this will have shifted up to age group 65+ years where 58 percent of the growth will have come from. In contrast, 46 percent of the growth in the poorest countries from 1955 to 1990 came at ages 0–14; from 1990 to 2025, it will be age the group 25–64 that contributes the same percentage, and only 30 percent will come at the childhood ages (Pool, 2000).

Looked at in more detail, one can trace the effects of population waves. This is illustrated in Figure 2 panels B and C, using Mexico and Thailand as case studies. Essentially, these show the progress of an 'ola' across significant life-cycle stages. In Figure 3, a slower, more dampened wave effect is shown for India with very modest double or multiple oscillations.

There are other countries which, because of their fertility histories, have multiple-oscillations. An example is my own country – New Zealand. Between the world wars TFRs dropped almost to replacement, followed by an extreme and prolonged baby-boom[3] – one exceeding those of all other developed countries. TFRs reached 4.5 around 1960, but more importantly birth cohort sizes were bimodal, peaking at almost the same numbers in 1960 and 1970. A baby-bust to below replacement in the late 1970s was followed around 1990 by a reprise that I have termed a 'baby-blip'. At this time rates reached back just to replacement (joining the United States and Sweden in this), and cohort sizes peaked at what was again almost equal to the sizes reached for the baby-boom's twin maxima. These fluctuations have projected a series of waves into the age pyramid, and these will hit various parts of the life cycle at different times and with varying impacts. For example, between 2006 and 2016 age groups 20–29, 50–59 and 60–69 years will be important drivers of overall growth, but will then decline sharply in significance being replaced by the 30–39 age group, and the older ages (70+). This means that services put in place to meet peak demands will then have to be reduced and the service focus shifted to meet the needs of another age-group (Pool 1999). The United States, shown in Figure 3B, has followed a similar trajectory to New Zealand's.

The Final Phase of the Age-structural Transition: Ageing

I will not discuss this phase in as much detail, simply because it is the last phase of the age-structural transition – curiously, the phase that is yet to come for most populations – that has gained the most attention analytically, theoretically, methodologically and in policy terms. Many of the key parameters have been identified statistically and this knowledge-base fed into policy (e.g. Stolnitz 1992, 1994).

At Cairo, with six percent of the world's population, the aged were given a special section of the final document, and ageing was seen as a priority issue even in Africa. By contrast, save for their sexuality, the 20 percent of the world's population aged 15–24 in 1990 were almost not mentioned. Their mobility, their labour force problems, even their contribution to parenting passed largely unnoticed, except for a call for them to be assimilated into society (United Nations 1995).

To illustrate the phase of ageing I use two very different case study countries. The Russian Federation is faced not only with the onset of ageing but multiple oscillations (Figure 3C) whereas France's ageing is much calmer (Figure 3D), for reasons Chesnais has already spelt out (Chesnais 1990: 335).

It is necessary at this point to reiterate a question raised earlier. The percent distribution in years beyond about 2010 is totally dependent on fertility assumptions, and, most notably,

[3] Rates were dropping in the 1960s with a minor secondary peak about 1970. But the cohort sizes, and the patterns of reproductive behaviour of baby-boom continued into the early 1970s, then changed radically from about 1975.

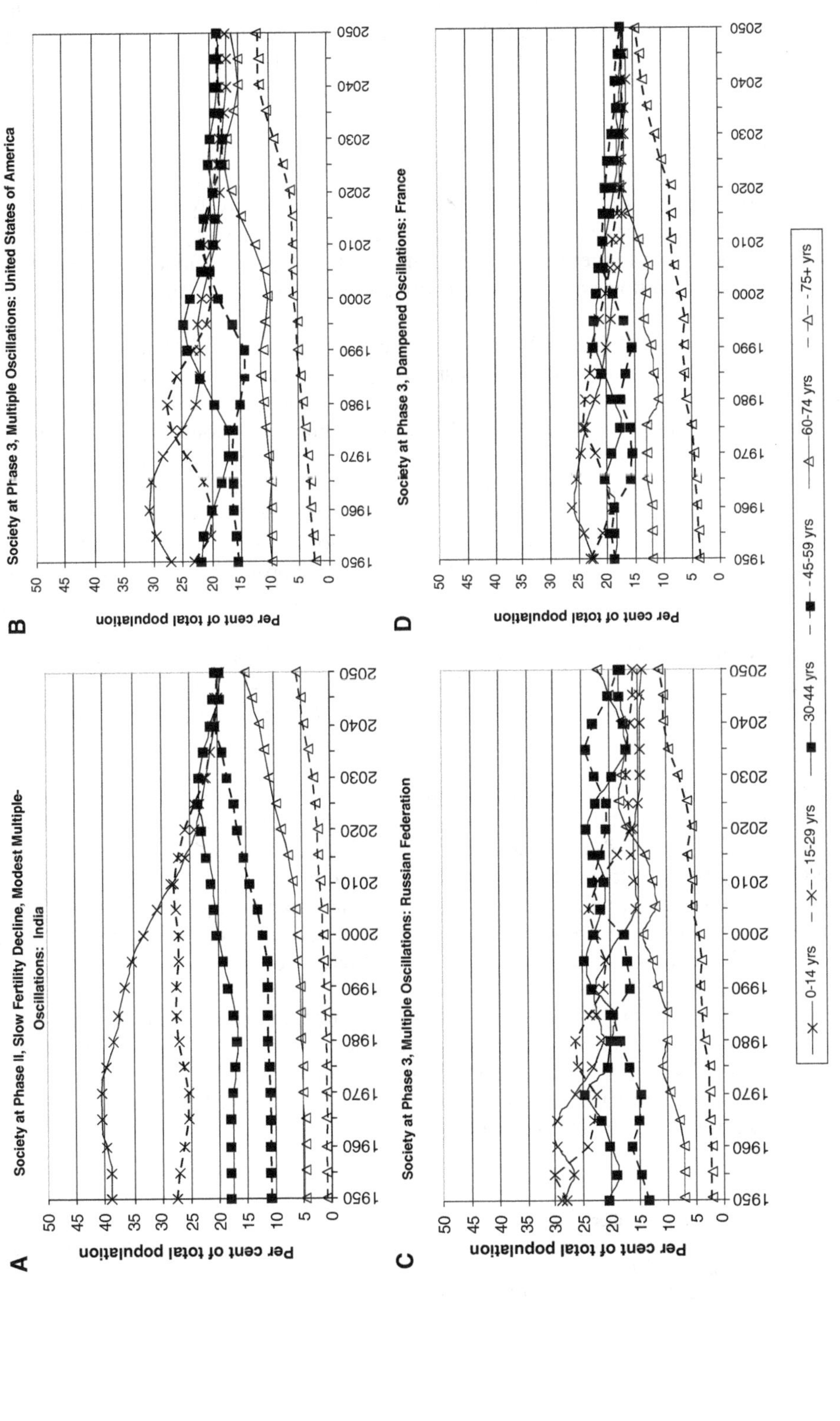

Figure 3. Percentage age distribution, functional age groups in India, United States, Russian Federation and France.

the retention of constant rates beyond a given reference year. Thus the oscillations seen in Figures 2 and 3 for future years, especially for the Russian Federation, are purely the effect of known existing cohort flows. The figures in the next section of the paper based more on measuring momentum effects are less compromised by this problem.

Cohort Flows

This last point is picked up in the data in Figures 4 and 5, that look at the way in which momentum produced by cohort flows – whether disordered or relatively well ordered – has an impact on particular life-cycle stages. For younger ages the results graphed in panel 1 for each case study is no longer robust from about 2010–15; but for older age groups the data on these numerical flows can be treated with a greater degree of confidence.[4]

In the case of the Congo (Figure 4A) the 'tidal wave' effect is being felt now at younger ages. The impact of civil war in the 1960s shows up quite dramatically for all ages over 0–14 years, and even as an echo-effect on the youngest cohort. Thus part of the decline in the numbers at 0–14 years reflects the effects of assumptions about fertility declines, and the impacts of the diminished size of the parental generation since about 1990. At older ages, and into the future until about 40 years from now, the quasi-stability achieved between the end of the 1960s and 1990 affects future trends. It must be stressed that while an attempt is made in the United Nations projections to show the future effects of AIDs, its full impacts will not have been clear when these projection series were being computed.

Mexico and Thailand (Figures 4B and C) are more advanced than Congo as examples of populations facing oscillations. There is a major wave and ebb coming from a rapid fertility decline, and rather muted secondary waves; the levels of their primary waves are also lower than the Congo's.

In China's case we see the results of a turbulent history[5] and not just of fertility decline in the 1970s. There was civil war and recovery in the 1950s, famine around 1960 and a subsequent recuperation, before a rapid decrease in fertility. This was followed by the consequences of the one-child policy accompanied by a mini baby-boom in the early 1980s, when age-at-marriage laws were introduced. Together these trends have had a major impact on cohort flows that have produced severe mutations. As a result, China faces major problems of policy formulation and implementation, a point I will return to later.

India (Figure 5A) makes an interesting contrast with China. Its natural increase and age-structural transitions have been more gradual and ordered, and waves are lower in level. As a consequence, India faces a demographically more ordered policy environment than does China; albeit, as Tuljapurkar and Li show (this volume, Chapter 5), the policy effects of age-structural transition are still marked.

---

[4] Obviously, total population numbers will also be affected by the reproductive trends of parental generations which are still at pre-parenting ages (or not even born themselves). But the effects of this will be more muted than for the results in Figures 1–3.

[5] This was detailed by Yan Hao in a paper on China to the Phuket, Thailand, meeting.

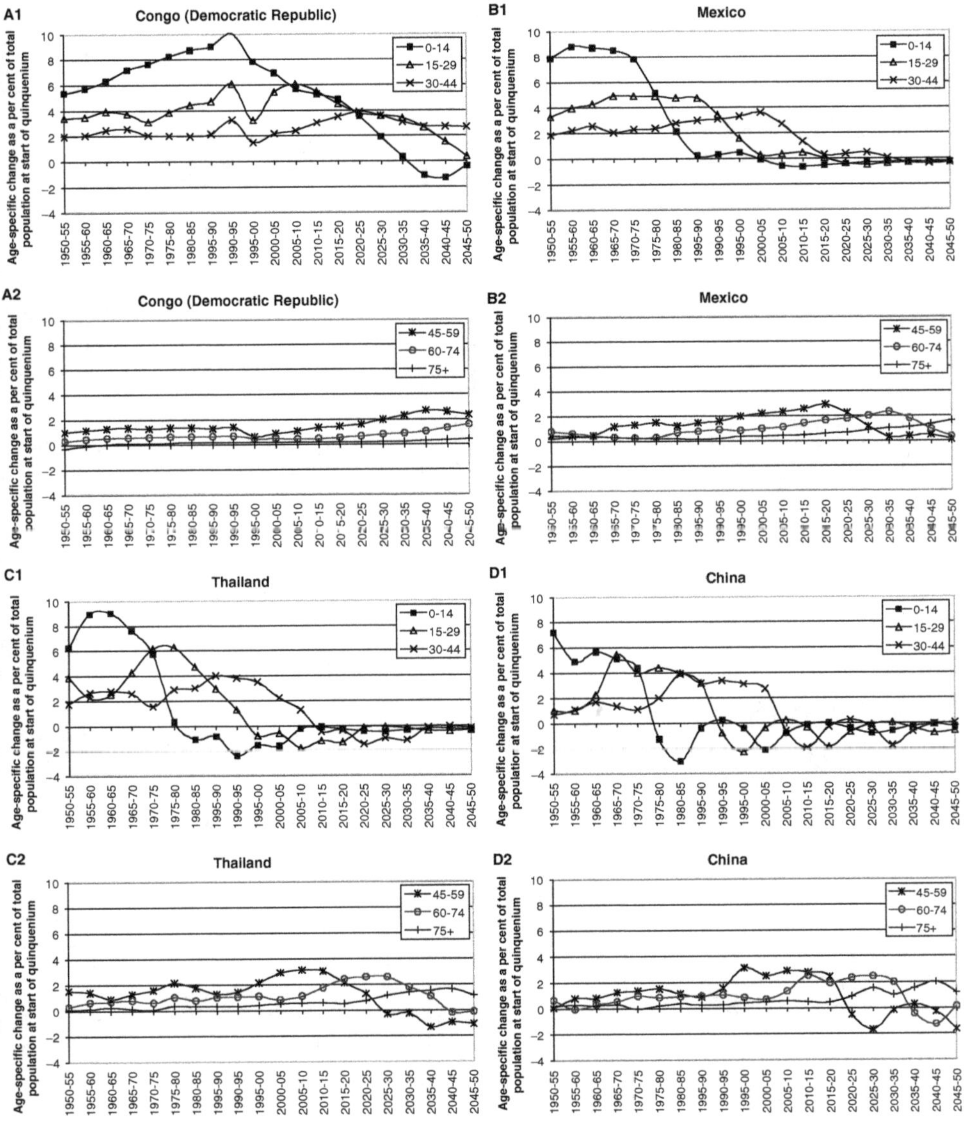

Figure 4. Impacts of cohort flows on age groups in Congo, Mexico, Thailand and China.

Turning to the category termed 'more developed countries', wave levels are lower. Nevertheless, the United States (Figure 5B), because of its recent fertility history, faces multiple oscillations.

Because of a long history of civil war and internal strife, the impact of the Second World War and subsequent shifts in patterns of natural increase, the Russian Federation

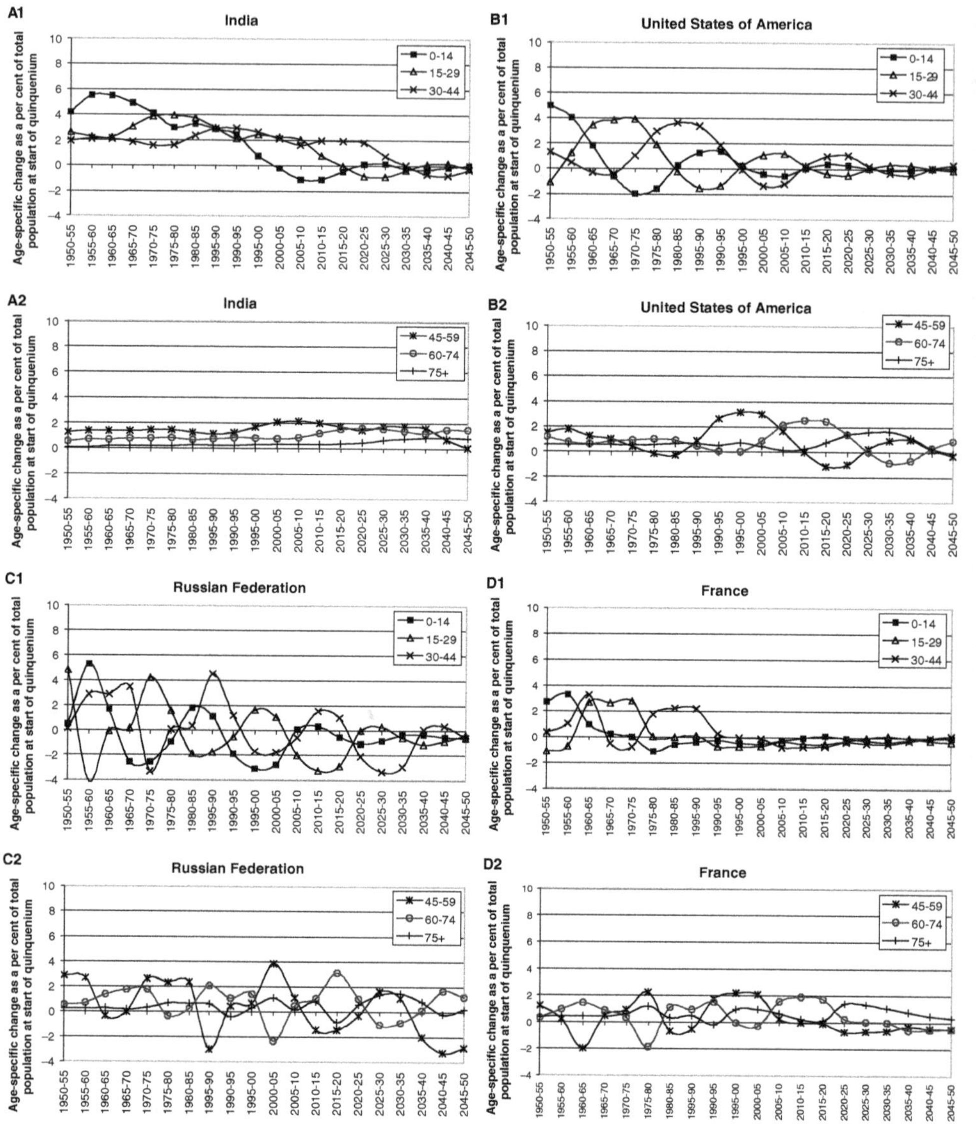

Figure 5. Impacts of cohort flows on age groups in India, United States, Russian Federation and France.

shows the most severe mutations of any of the more developed country examples. There is an extreme series of waves and disordered cohort flows in all years shown in Figure 5C.

One can compare the United States and the Russian Federation with the case of France shown in Figure 5D. The latter's waves are low and relatively ordered.

Figures 1–5 have described different sorts of age-structural transitions and their attendant cohort flows. As might be expected the more muted the natural increase regimes (fluctuations and trends in levels of fertility and mortality) the less marked the subsequent oscillatory effects. But it is the interaction between these that seems crucial. Thus, Mexico faces a marked 'ola' effect, but not multiple oscillations; for the United States, wave levels are not high, but there are significant levels of disorder. The Russian Federation is highly disordered. In contrast, India and France benefit from relatively ordered age-structural transitions, France particularly so (Pool, in press).

4. Age-Structural Transitions and Policy: Background

Much of the literature on population and development interrelationships has tended to focus on population growth overall (e.g. Simmons 1988), or on the growth of very wide age groups (e.g. the 'working ages'). To the extent that it dealt with development at all, the Cairo Conference in 1994 certainly had that emphasis (United Nations 1995: especially Chap. 3 that in spite of its title does not elaborate structural issues).[6]

Most studies have not attempted systematically to take account of the policy implications of age-structural transitions, although the situation is starting to change. For example, a relatively recently edited, seminal collection of papers has taken a wider perspective in a number of its chapters (Cassen et al. 1994: especially chapters by Cassen himself and Thomas Merrick). Another example is in a study on "Age structure effects and growth in the OECD, 1950–1990" (Lindh and Malmberg 1999; for the US, see Fair and Dominguez 1991). Of particular interest because it is so innovative, and draws relationships between population and the fiscal and financial sectors, is the work on savings and investment in the Asian region, of which the work by Higgins and Williamson (1997) is a very pertinent example. Using relatively simple demographic indices they show how the age structure is related to needs for external capital. Finally, as noted already, a partial exception would be in the area of ageing, around which a significant literature has grown (e.g. Stolnitz 1992, 1994; Rowland 1991, to cite only three of many examples).

It will be noted that these examples are mainly drawn from developed countries. Yet the analysis earlier in this paper showed that momentum effects and the mutations produced by age-structural transitions are severe in the Third World. Of critical importance for the analysis of the interface between population and development/policy in the developing world is the way in which age-structural transitions are manifested by three phenomena identified earlier: accelerating momentum, changes in momentum, and oscillations produced by disordered cohort flows. The first carries with it the implication of exponential increases in demand for services, and the second and the third to fluctuating demand.

Whichever trajectory is being followed, age-structural change will put differential pressures on key life-cycle stages, and thus on policies responding to the needs of populations at those

---

[6] In Vallin's (1994) words "this conference on population and development hardly ever dealt with development" (present author's loose translation). See also Loriaux 1994.

ages. An example of the importance of this issue comes in a paper describing a developed country, and dealing with changes that in quantum terms are far less marked than has been the case for Mexico, which was used as an example earlier in this paper. This is the United States, and the example relates to one of the most sensitive sets of relationships between age structure and employment, in the youth labour market, and the way that birth cohort size can affect the "fortunes of one's birth" (Macunovich 1999). These interrelationships are complicated by the fact that they are likely to occur in an 'on-again/off-again' way. (Chapters 3, 4 and 11 by Jones, Xenos, and Urich and Gultiano address this issue in more detail.)

This point is taken further in Figures 6 and 7 using two age groups with high 'demographic densities'. Rindfuss (1991) used this term to describe the numerous biological, social, cultural, psychological and other changes – to say nothing of finishing education and taking up a job – which an individual goes through over the teenage and early adult years. I have extended this to a later stage in the life cycle, about the time retirement is reached at, say, 60–74 years. To standardise this, the percentage at these ages in the case studies is compared with the proportion globally at the same ages.

Development and social policy initiatives are frequently required to respond to these changes. Thus, if an exceptionally large cohort reaches these key life-cycle stages, then the policy and fiscal demands (or their analogues within families) will be maximised. At one extreme is the Congo (Figure 6A) where demands on sectors such as the labour market will continue unabated and increase exponentially. India (Figure 7A) will see this continue, but in a more absorbable way and Mexico (Figure 6B) to a lesser degree. Thailand (Figure 6C) and China (Figure 6D) will see abrupt shifts from demands at youth ages to those for early retirement. In the more developed countries (Figures 7B, 7C and 7D) pressures will be on policies and services for older people, but there is variation in the degree to which demands will fluctuate, as well as in the levels relative to the world as whole. Again it is the Russian Federation that is most extreme both in levels and its fluctuations.

In the last section of my chapter I will attempt to identify some of the policy implications of age-structural transitions. Underlying this is the argument that a focus on 'age-structural changes and development' is a far more effective strategy than the use of less refined 'population and development' models. The analysis of the interrelationships between age-structural transitions and development provides a knowledge-base that is more focused, yet flexible. It is thus more appropriate for approaches to development policy that emphasise strategic approaches, and rolling and sectoral planning (Siwatibau 1993).

5. Age-Structural Transitions and Policy: Towards a Framework

What might be the central features of a framework? Here I draw on and develop an earlier analysis in which some of the basic parameters were laid down (Pool 1994).

A starting point is that most social behaviours, even if they are mediated by factors such as culture or personal life-style, are age-specific. This includes almost all demands for services dealt with by public policy, as well as patterns of consumption, including

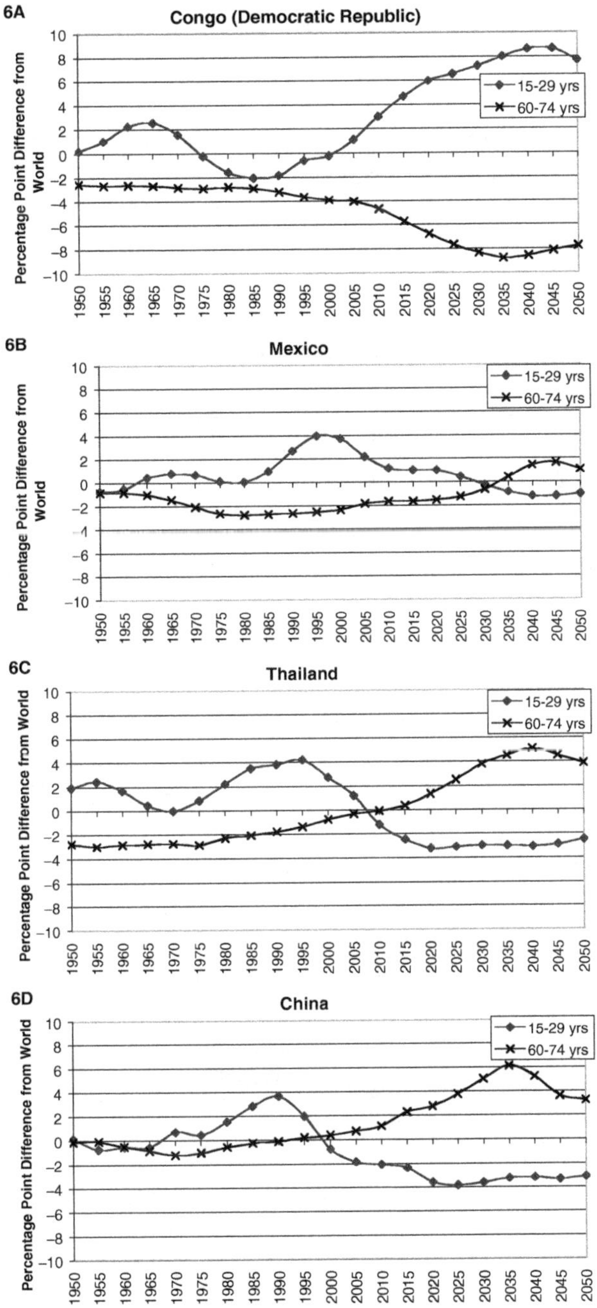

*The term used by Rindfuss (1991).

Figure 6. Impact of age-structural transition wave effects on two age groups subject to high Levels of "demographic density".*

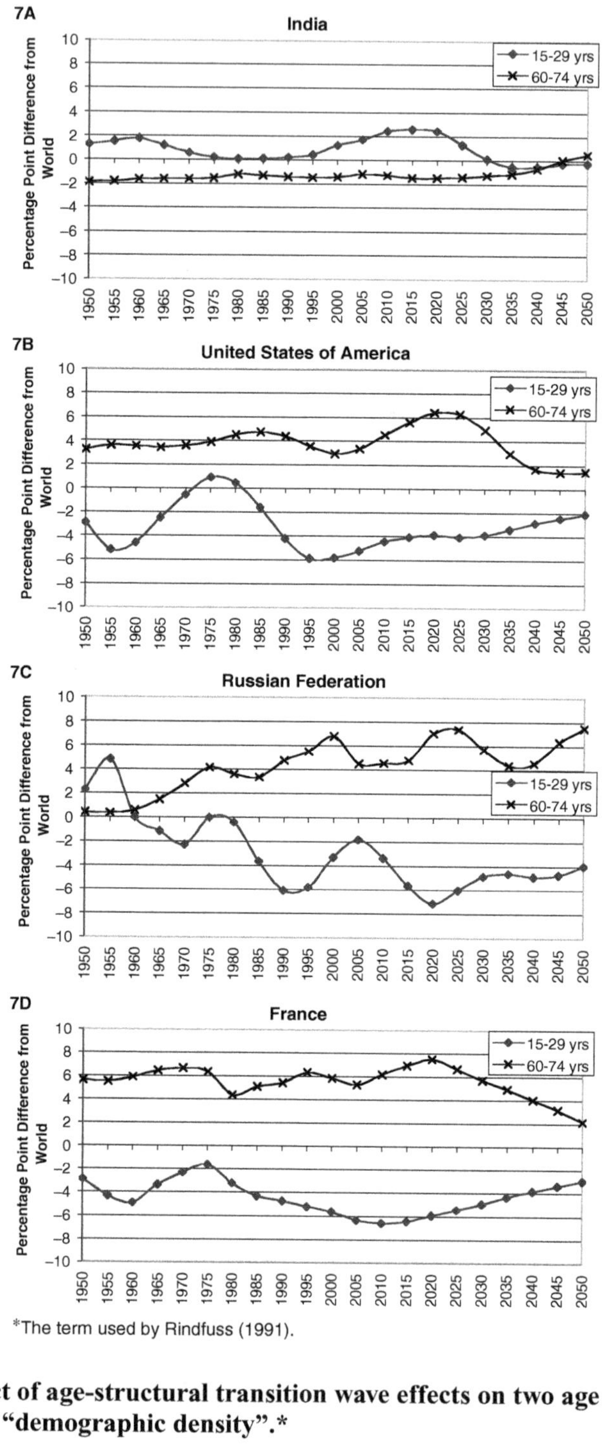

*The term used by Rindfuss (1991).

Figure 7. Impact of age-structural transition wave effects on two age groups subject to high levels of "demographic density".*

discretionary expenditure. Marketing people recognise this relationship as they attempt to segment markets, but for many public sector domains this is less frequently taken into account.

To a considerable degree, population change drives the demand-side of development – the need for goods or services, and the fiscal burdens implicit in meeting those demands: who needs what tax-based support. But there are also supply-side effects, the workforce being an obvious example of this. In countries subject to extreme momentum effects or to large waves, demographic supply will typically outrun demand for workers. In countries facing troughs at pivotal ages, the demand for workers in key industries will often outrun supply. This raises an important point to be developed below: the inter-linkages between age-structural changes and sectoral or industrial development. A less evident but certainly important supply-side issue is the question of fiscal capacity: who pays the taxes.

Fiscal burdens and capacities are not the only manifestations of the relationship between age-structural changes and development. In many societies, a high proportion of the workforce and the economy will be outside the formal sector, or even outside the market. Yet these populations have burdens and must meet needs analogous to those covered through the fiscal system in market economies. The difference is that it is the family that plays a key role in these societies, shouldering the burdens and fulfilling capacities. But even in developed countries with market economies the family meets many needs. For example, a 'free' education system rarely if ever covers all the costs attendant on schooling children. At the very least the family probably pays for clothing and some school supplies.

It is for this reason, therefore, that in reviewing age-structural transitions it is essential to identify their interface with family/household structural transitions. Indeed, family structures passing through transitions will also typically be undergoing their own age-structural transitions that have both macro- and micro-level impacts. The shift from precocious to delayed childbearing in developed countries, especially in North America and Australasia, is an obvious example. At a micro-level, this has entailed a move to later ages of every single family function, especially those relating to the nurturing process. The legislating of a legal age at marriage in China in the early 1980s and the rush of couples, who previously would have been subject to community sanctions favouring later marriage, to marry younger, produced a major wave (Yan Hao's paper to the Phuket meeting November 2000). Beyond this, age-structural transitions are themselves the drivers of changes in family formation and structures. Birth cohort size, for example, is the product of the age-specific fertility rate and the female age-structure within the reproductive span. In backblocks of New Zealand in the 1880s, when the national Total Fertility Rate was above 6.0, marital General Fertility Rates exceeded 250–300 per 1000 women aged 15–49 years, simply because of composition effects. The pioneer families consisted almost entirely of young couples in their twenties at prime reproductive ages (Pool and Tiong 1991). In developed countries the weighting of the female population of reproductive ages towards late ages will place constraints on policies attempting to achieve replacement and to increase birth cohort sizes in order to offset the effects of ageing, even if age-specific rates rise (Pool, in press).

As was shown in the earlier part of this paper, a central age-structural issue is that cohorts of varying sizes pass through age groups, or life-cycle stages, placing demands on services and pressures on patterns of consumption typifying those age groups. Equally well, the supply of services will be in part a function of the number of persons who are at ages at which the workers providing various services will be found. This inter-dependency between age and the demand and supply of services is central to the population/development relationship.

These points raise another important issue. Because demographic change plays such a major role in driving demand for goods and services, there is a tendency to view population and public policy as a component of social policy. This is true even in the debates on pensions, where age-structural data are used for the demand-side, but fiscal and financial forecasts typically use economic scenarios that exclude the human capital factors, and thus do not cover the demographic drivers affecting fiscal capacity. Yet because of their role in terms of fiscal (or familial) burden and capacity, age-structural changes are as important for the fiscal and financial sectors as they are for the more clearly defined social policy sectors. If this had ever been in doubt, then the recent work on Asian savings and investment referred to above (e.g. Higgins and Williamson 1997) would dispel it. This point is of major practical significance for population and development as typically in the past the endogenisation of population into policy and planning has seldom been well developed in those ministries dealing primarily with fiscal and financial concerns. Where this has been attempted it has often faltered because the global population changes that will unfold over 10 years or longer will not be seen to have a great impact on the economic trends that are being projected over the normal economic planning time horizon of three to five years (Lindh and Malmberg 1999; this point was also stressed but is less well demonstrated statistically in UNFPA 1989).[7]

It is clear that much of the inter-relationship between population and development can be refined and perhaps made more effective in terms of outcomes by adopting strategies that identify sectors which are primarily concerned with one major life-cycle stage. The impacts of both demand and the human capital necessary to supply services can be built into such analyses. Improved modelling of age-structural transitions could then be used to project when peak demands for expenditure and services will occur. Such a model exists already, and is reasonably well developed, for one particular sector: basic education.[8] Generally, the accent would be placed on normative needs and services for populations at various life-cycle stages. Such a strategy can be carried across to deal with social problem areas, such as need for welfare, pressure on prison services, etc. Table 3 identifies some key sectors that deal with primary universal needs in specific life-cycle policy domains. Not on the table are the fiscal and financial sectors that perform a cross-cutting role. There are other sectors such as transport, communications, sewage and water, or energy for which age-structural changes are of some importance as they affect patterns of

---

[7] One study was based on a global evaluation. The present author was a member of the evaluation team that prepared that report and also regional reports on Asia-Pacific and Sub-Saharan Africa.

[8] Tuljapurkar and Li model this for India: See Chapter 5.

Table 3. Services meeting primary universal needs for persons at major life-cycle stages.

Life-cycle stage				
Childhood	**Youth**	**Middle ages**	**Older ages**	**Very old age**
(0–14 yrs)	(15–24 yrs)	(25–64 yrs)	(65+ yrs)	(80+ yrs)
Health and nutrition	MCH/FP	MCH/FP	Health	Health
Education	Education	Employment	Employment	
		Housing	Housing	Institutionalised housing
		Income	Income	Income maintenance

Note: MCH = Maternal and Child Health. FP = Family Planning.

demand, but where the population-development link is more likely to be driven by overall growth.

Table 3 relates to universal needs and thus has a connotation of public policy rather than of market-based services. Nevertheless, the general point raised above also relates to the provision of private sector goods and services, including those that depend on discretionary expenditure. Moreover, in numerous sectors the intervention of private enterprise is very much a public policy concern, especially in those countries, perhaps the overwhelming majority, where the economy is mixed.

Finally, in looking at age-structure and policy, it is necessary to recognise the effects of inter-generational linkages and dependencies, and competition for resources. The first has been relatively well rehearsed in the area of ageing in terms of support systems. This also applies to transactions between parents and dependent children.

The notion of competition has also been raised in the context of ageing in reference to generational accounting. The focus has essentially been on savings for old age, and whether or not one generation is being taxed to support the retirement of another. But this issue is far more complex than most generational accounting exercises might suggest. In countries subject to double- or multiple-oscillations there will inevitably be competition for resources between different cohorts, at whatever life-cycle stage they are at. This is seen for a number of the countries presented in Figures 4 and 5.

In New Zealand or the United States, the baby-boom demanded the significant weighting of expenditure towards paediatric health and educational services, and family housing. Moreover, even to develop the human capital necessary to support the retirement of the baby-boom cohorts may require heavy expenditure today on cohorts currently at childhood ages, particularly in those countries like New Zealand that had an upsurge in birth cohort sizes around 1990. Viewed from one perspective this could be a bonus, as the baby-blip (echo to the baby-boom) provides a population base that could offset the increases in the old age dependency burdens produced by the ageing baby-boom cohorts. But equally well it requires an investment in sectors such as education and employment if the potential of the baby-blip is to be realised (Pool 1999).

6. Towards a Synthesis

This paper does not lend itself to grand concluding statements. Rather it is necessary to end by arguing that it suggests a need to set a new agenda for demography. The world is shifting demographically from change driven by growth *per se*, towards drivers coming from complex restructuring by age and cohort. There is an urgent need, then, to turn our attention to these structural changes, especially relating to age groups and to cohorts. The fact that these structural changes will often be turbulent or disordered makes the intellectual and policy issues more challenging. Indeed it is an agenda that demands the attention of the very best minds in the discipline if we are to develop the methodological and theoretical tools that permit the refined analysis of these issues.

We have an existing mobilisation model. The Princeton Africa project not only established that it was possible to carry out serious demographic research on Sub-Saharan African topics, but it also launched several decades of intellectual endeavour that greatly enriched the field, particularly in the area of the methodologies of indirect estimation of vital trends. There is now a need to mobilise the field through new initiatives that meet the emerging challenges facing us in the structurally turbulent decades ahead.

Acknowledgements

The New Zealand Foundation for Research Science and Technology supported the research on which this paper has been built. I wish to thank Dr A. Dharmalingam who reviewed an earlier version of the paper and provided valuable critical suggestions.

References

Cassen, R. and Contributors. (1994), *Population and Development : Old Debate, New Conclusion*, Overseas Development Council: Washington, DC.

Chenery, H. and Syrquin, M. (1975), *Patterns of Development*, 1950–1970. Oxford: Clarendon.

Chesnais, J-C. (1986), "La Transition Demographique: Etapes, Formes, Implications Economiques," *Population*, 41(6):1059–70.

Chesnais, J-C. (1990), "Demographic transition patterns and their impact on the age-structure." *Population and Development Review*, 16(2):327–54.

Coale, A. (1972), *The Growth and Structure of Human Populations*, Princeton University Press: Princeton, NJ.

Collomb, P. et al. (1996), "Besoins alimentaires et croissance démographique." *Sommet Mondial de L'Alimentation: Documents d'Information Technique*, 1. FAO: Rome.

Crosby, A.W. (1986), *Ecological Imperialism: The Biological Expansion of Europe, 900–1900*, Cambridge University Press: Cambridge.

Demeny, P. (1997), "Replacement level fertility: the implausible endpoint of the demographic transition." In: *The Continuing Demographic Transition,* G., Jones, R., Douglas, J. Caldwell and R., d'Souza, eds. pp. 94–110. Oxford: Clarendon.

Dittgen, A. (2000), "Apres la transition demographique: l'équilibre ou les turbulences?" *Paper, Colloque de l'AIDELF "Vivre plus longtemps, avoir moins d'enfants: Quelles implications?". Byblos, 10–13 Oct.*, Séance 4. pp. 19–26.

Drewnoski, A. and Popkin, B. (1997), "The nutrition transition: new trends in the global diet." *Nutrition Reviews*, 55(2):31–43.

Dyson, T. and Murphy, M. (1985), "The onset of fertility transition." *Population and Development Review*, 11(3):399–440.

Fair, R. and Dominguez, K. (1991), "Effects of the changing US age distribution on macro-economic equations." *American Economic Review*, 81, 5:1276–94.

Frejka, T. (1982), "Momentum." In: *International Encyclopaedia of Population*, Vol. 2. J. Ross, ed. pp. 450–451. Free Press: New York.

Higgins, M. and Williamson, J. (1997), "Age structure and dependence on foreign capital." *Population and Development Review*, 261–93.

Horiuchi, S. (1995), "The cohort approach to population growth: a retrospective decomposition of growth rates for Sweden." *Population Studies*, 49:147–63.

Jones, G. (2005), "Human Capital Aspects of Economic Development: A Comparative Perspective in Asia." *Population, Resources and Development*, Springer, Dordrecht, pp. 41–56.

Jones, G., and Douglas, R. (1997), "Introduction." In: *The Continuing Demographic Transition*, G. Jones, R. Douglas, J. Caldwell, and R. D'Souza, eds. pp. 1–12. Clarendon: Oxford.

Keyfitz, N. (1968), *Introduction to the Mathematics of Population*, Adison-Wesley: Reading, MA.

Lepina, M-C. and Pool, I. (2000), "Social and Spatial Differentials in Age-structural Transitions: Implications for Policies in Ageing." *Population, Espace et Societé*, 399–410.

Lévy, M. (1998), "Raisonner sur le vieillissement." *Population & Sociétés, Bulletin Mensuel D'Information de L'Institut National D'Etudes Démographiques, Numéro 341, Dec.*

Lindh, T. and Malmberg, B. (1999), "Age-structure effects and growth in the OECD, 1950–1990." *Population Economics*, 12(3):431–50.

Loriaux, M. (1990), "Il sera une fois... la révolution grise jeu et enjeux d'une grande mutation sociale." In: *Populations Agées et Révolution Grise*, M. Loriaux, D. Remy, and E. Vilquin, eds. pp. 3–32. Editions Ciaco: Bruxelles.

Loriaux, M. (1994), "Les mystères de la Grande Pyramids." *La Lettre du CIDEP, Louvain-la-Neuve, Oct–Dec.*

Macunovich, D. (1999), "The fortunes of one's birth: relative cohort size and the youth labour market in the United States." *Population Economics*, 12(2):215–72.

Maddison, A. (1982), *Phases of Capitalist Development*, Oxford University Press: New York.

New Shorter Oxford English Dictionary, (1993), 2 volumes. Clarendon: Oxford.

Notestein, F. (1945), "Population: the long view." In: *Food for the World*, T. W. Schultz, ed. Chicago University Press: Chicago, IL.

Omran, A. (1982), "Epidemiologic transition." In: *International Encyclopaedia of Population*, Vol. 1. J. Ross, ed. Free Press: New York.

Pool, I. (1991), *Te Iwi Maori: A New Zealand Population Past, Present, Projected*, Auckland University Press: Auckland.

Pool, I. (1994), *The World's Caring Capacity : Population Compositional Aspects*, Ind. Commission Pop and Quality of Life: Paris.

Pool, I. (1999), "People (= population) and public policy in New Zealand." *New Zealand Population Review*, 25:57–79.

Pool, I. "Vers un modele de la 'transition age-structurelle': une consequence mais aussi une composante de la transition demographique." Paper presented at *Conference of Association Canadienne Française pour l'Avancement de Science, Montreal, May*, 2000.

Pool, I. "Age structural transitions in industrialized countries." Paper presented at *IUSSP and NUPRI Seminar, Mar. 2001*, to appear in *Responses to Aging in Advanced Industrial States: Riding the Age Waves.* Shripad Tuljapurkar, Naohiro Ogawa, eds. IUSSP series on population.

Pool, I. and Tiong, F. (1991), "Sub-national differences in the Pakeha fertility decline, 1876–1901." *New Zealand Population Review*, 17, 2:46–64.

Preston, S. (1986), "The Relation between actual and intrinsic growth rates." *Population Studies*, 40:343–352.

Preston, S. (1988), "Reply to Wachter." *Population Studies*, 42:495–501.

Preston, S. and Coale, A. (1982), "Age structure, growth, attrition and accession: a new synthesis." *Population Index*, 48:217–59.

Rindfuss, R. (1991), "The young adult years: diversity, structural change and fertility." *Demography*, 28(4):493–512.

Rowland, D. (1991), *Population Ageing in Australia, Paris and Valletta*, CICRED and INIA: Paris.

Rowland, D. (1996), "Population momentum as a measure of Ageing." *European Journal of Population*, 12:41–61.

Shryock, H., Siegel, J. et al. (1976), *The Methods and Materials of Demography*, Academic Press: New York.

Simmons, O. (1988), *Perspectives on Development and Population Growth in the Third World*, Plenum Press: New York.

Siwitibau, S. (1993), "Macro-economic Paths and Population in Pacific Island Countries." Paper presented at *Meeting of Senior Officials on Population and Sustainable Development, sub-regional meeting prior to Cairo Conference, Port Vila (Vanuatu), Sept.*

Stolnitz, G., ed. (1992), *Demographic Causes and Economic Consequences of Population Ageing*, New York University: New York.

Stolnitz, G., ed. (1994), *Social Aspects and Country Reviews of Population Ageing*, New York University: New York.

UNFPA. (1989), *Comparative Evaluation of UNFPA Support to Population and Development Planning: Global Report*, UNFPA: New York.

United Nations. (1992), *Sex and Age Distribution of World Populations*, New York University: New York.

United Nations. (1995), *Population and Development: Programme of Action Adopted at the ICPD, Cairo, Sept. 1994*, Department for Economic and Social Information and Policy Analysis: New York.

United Nations (Population Division). (1992, 1998), *World Population Prospects: the 1992 Revision*, New York University: New York.

United Nations. (2000), *World Population Prospects. Volume 1: Comprehensive Tables*, New York University: New York.

Urich, P. and Gultiano, C. (2000), Paper to *Phuket meeting, November 2000.*

Vallin, J. (1994), *La Chronique du CEPED*, automne.

van de Kaa, D. (1987), "Europe's Second Demographic Transition." *Population Bulletin*, 42, 1. Population Reference Bureau: Washington, DC.

van de Kaa, D. (1988), *The Second Demographic Transition Revisited : Theories and Expectations*, Neth. Inst. for Advanced Study in the Soc. Sci: Werkstukken, Wassenaar.

Yan H. (2000), Paper to *Phuket meeting, Nov. 2000.*

Wachter, K. (1988), "Age group growth rates and population momentum." *Population Studies*, 42:487–92.

Xenos, P. (2000), Paper to *Phuket meeting, Nov.*

Zelinsky, W. (1971), "The hypothesis of the mobility transition." *Geographical Review*, 61(2): 219–49.

CHAPTER 3. HUMAN CAPITAL ASPECTS OF ECONOMIC DEVELOPMENT: A COMPARATIVE PERSPECTIVE IN ASIA

GAVIN W. JONES
National University of Singapore

Introduction

This paper will deal mainly with those Asian countries that are in the later transitional phase rather than the early phase of demographic transition. I take the later transitional phase to mean the phase when mortality has already fallen to moderate levels (infant mortality rate of 50 or less) and total fertility rate has also fallen below 3. In other words, considerable transition has already occurred, with significant effects on age structure, but not to the point where the main issues are ageing and fertility levels far below replacement.

The paper will argue that human resource development has had much to do with both the demographic transition in these countries and their generally rapid rates of economic growth. It will show that declining fertility affects numbers entering the school-going ages and later the working ages. Fertility decline both eases the burden of expanding education systems and lowers dependency burdens over a considerable period of time. But rapid expansion of education systems can beget more problems, if job opportunities are not increasing rapidly enough to absorb the growing numbers of educated young people entering the workforce. Finally, the paper shows the importance of planning ahead for fluctuations in the growth of particular age groups, caused by rapid declines in fertility and, in some cases, echo effects resulting from temporary increases in fertility.

If we examine the current demographic situation of the countries of Asia, there is of course great variety. The infant mortality rate (IMR) and total fertility rate (TFR) for the different regions in 1995–2000 were as follows:

South-Central Asia: 73 and 3.36
South-East Asia: 46 and 2.69
East Asia: 38 and 1.77

South-East Asia epitomises the later transitional phase, in contrast to South-Central Asia (especially when the Central Asian countries are excluded), which is placed between the emergent transitions and the later transitional phase; and also in contrast to East Asia, which is very much into the post-transitional phase. For this reason, and also because of the author's greater knowledge of this region, this paper will emphasize South-East Asian

S. Tuljapurkar et al. (eds.), Population, Resources and Development, 41–56.

countries, but will also use Iran as an interesting case study of the impacts of fluctuating fertility levels.

The eleven South-East Asian countries are not homogeneously 'later transitional'. Only seven had IMRs of less than 50 in 1995–2000 and only six had TFRs of less than 3. However, the countries with the higher mortality and fertility levels were mostly very small ones, except for the Philippines, which had a relatively high fertility level; as a result the population-weighted South-East Asian figure is definitely 'later transitional'.

The Economic Development Record of East and South-East Asia and the Likely Role of Human Resource Development

Though the performance of some Asia-Pacific economies received a severe setback as a result of the economic crisis from 1997 on, up to that point these economies had grown very strongly over three decades. The causes of this strong growth performance have been much debated. It seems to be generally agreed that the causes must be embedded somewhere in the following list of factors: (i) strong investment in infrastructure; (ii) an efficient absorption of advanced technology; (iii) a stable political environment; and (iv) an impressive commitment to human capital formation (Ogawa, Jones and Williamson 1993: 2). These countries had a long-standing and strong emphasis on educational development. A number of them have demonstrated far steeper increases in their commitment to educational investment, compared with all the Western industrialized nations in the nineteenth century and almost all the contemporary developing countries in Latin America and Africa. The efficacy of this commitment to education has been further enhanced by the decline in dependency ratios consequent on fertility declines.

It is difficult, of course, to prove the impact of human capital deepening (e.g. the increase in average numbers of years of schooling of workers) on the economic growth rates of these countries. But the theoretical linkages between a strong commitment to educational investment and rapid economic growth are emphasized by the new endogenous growth theories, which endogenize the rate of technological advance through human capital accumulation, most notably through formal schooling or informal skill development on the job, and some of which make the rate of human capital accumulation an increasing function of the level of human capital. There is considerable empirical support for these theories in inter-country growth studies (e.g. Barro 1991).

Declining dependency ratios and strong investment in human capital are inter-related; reduced dependency ratios have facilitated human capital deepening, particularly through an increase in public spending per secondary school student. These have interacted favourably in the Pacific Rim countries with their reasonably high prior levels of investment in human resource development. Declining youth dependency ratios over recent decades in South-East and, especially, East Asia have resulted in a rise in savings rates and decreasing foreign capital dependency (Higgins and Williamson 1997). This 'demographic bonus', which will be discussed in more detail below, persists over many decades and gives countries the chance to build strong and resilient economies well before significant increases in overall dependency – resulting from increased proportions of the elderly – occur.

Declining Fertility, Schooling Burdens and Educational Quality

One of the earliest impacts of declines in fertility is on numbers entering the primary-school-going ages. Whereas the passage of five years may lead to a 15 percent increase in numbers of annual entrants to the school-going cohort in a country with traditionally high fertility, a reasonably rapid decline in fertility can soon lead to a complete cessation in this growth. This has enormous implications for the task of reaching universal primary school education. Many countries in Asia have reached the point where the growth of their primary school-aged population ceased. The countries and the approximate years when this growth ceased are as follows:

Singapore	1968
Republic of Korea	1972
Hong Kong	1972
Taiwan	1972
China	1980
Thailand	1985
Indonesia	1987
Sri Lanka	1990
Myanmar	1990
Iran	1998

Interestingly, in both Singapore and Thailand, this cessation of growth came at roughly the same time as these countries reached universal primary education, though this point was reached about 17 years later in Thailand than in Singapore (Jones 1990: 36–40). In neither case did the cessation of growth of the denominator of their enrolment ratios slow the rise in educational expenditures; rather, it permitted these expenditures to achieve more in terms of expanding secondary education and increasing expenditures per pupil (and presumably, quality of education) at all levels. Schultz (1985) has demonstrated that larger school-age cohorts are associated with lower expenditure per pupil.

A cessation of growth of youth populations does not necessarily mean that growth will not resume at some later stage. We will deal with such 'echo effects' later. The key point here is that fertility declines lead to a cessation of growth of the school-age population, and in many cases a slow decline, for a considerable period of time. This is a very favourable outcome for provision of educational facilities, irrespective of the point countries have reached in their educational provision. If they have not yet attained universal primary education, the burden of doing so will be lightened. If they have already attained universal primary education, but the quality of this education leaves much to be desired, the task of raising quality will be ameliorated by the removal of the pressure to increase the number of school places. And if they are in the unusual situation of being satisfied with both the quantitative and qualitative situation of their primary school education, the resources freed by the cessation in growth of the primary-school-age population can be used to make more rapid gains at the more expensive secondary education level.

For countries such as Thailand and Indonesia, the key target in recent years has been to raise the transition rate from primary to lower secondary education. Thailand has succeeded in increasing this transition rate very rapidly over the 1990s. In contrast, Indonesia has struggled to raise the transition rate. It climbed to slightly over 80 percent in 1996, but declined again after the onset of the economic crisis in 1997. There is no doubt that the fact that numbers of school-age children were not increasing helped to maintain transition rates to secondary education in the face of the economic crisis.

Declining Fertility and Improvements in Health and Nutrition

Declining fertility in countries in the later phases of the demographic transition facilitates improvements in the health and nutrition of children, with important effects on productivity (Behrman et al. 1988). Three points that might be made in this context are that: (1) improvements in the nutritional status of children that result in increased adult heights can have substantial long-term pay-offs in terms of increased agricultural productivity and earnings of these children when they become adults; (2) improved nutritional status in the early ages affects children's subsequent schooling performance and future productivity as adults; (3) nutrition in the early ages significantly influences the productivity of investments in education and health. Much of the benefit of nutritional improvement appears to be long-term in nature. The synergy between fertility reduction, investments to expand the coverage of education, and investments in health and nutrition is obvious.

Declining Fertility, Educational Attainment of the Workforce and Dependency Burdens

We need to take a broad perspective on the demographic dynamics of the rise in educational attainment rates. From a 'human capital' perspective, the key benefit of raising the average educational attainment of labour force entrants is that this in turn raises the educational attainment of the labour force, thus facilitating more rapid economic development. One potential benefit of a population with high fertility is that the rate at which the existing labour force is being replaced is higher than in a population with lower fertility. As Keyfitz (1989) argued in relation to Indonesia, in such a population, there may be four workers entering the labour force for every older worker leaving. Importantly, there tends to be a steep gradient in educational attainment from the older labour force ages, those in their 50s and 60s, to those in their 40s, and in turn to those in their 30s, 20s and teens.

The more rapid entry of these better-educated younger workers into the labour force is very much to be desired. Leibenstein made this argument in favour of high fertility in 1971. What he failed to recognize, however, was that if the rapid increase in numbers at the school-going ages hindered the efforts to keep the average child in school longer, much of the benefit he rightly identified could be lost. There was an additional weakness in Leibenstein's argument. The very act of keeping children in school for a longer period so that they could enter the labour force with a much higher level of education than the older workers slows the process of replacing older workers with better-educated workers. The young people entering the labour force at ages 10–19, for example, are those who are not continuing further with their education. Thus the average educational attainment of young workers entering the labour force at ages 10–19 is lower than the average educational attainment of the same cohort when it reaches ages 20–29 (Jones and Gingrich 1968;

Jones 1976). Therefore, even if a higher-fertility population succeeded in raising school enrolment ratios as rapidly as a declining fertility population, this would make very little difference to the time taken to replace the poorly-educated labour force with a group of better-educated workers.

Another way of examining the economic benefits of age structure changes associated with demographic transition is to study trends in dependency ratios. Normally, these are calculated as the sum of the population aged 65 and above, together with those below 15, divided by the population aged 15–64 (normally times 100). The idea is to find the rough proportion of dependents to producers. Modifications can be made to allow for the extension of schooling beyond 14 for most young people in many countries, as well as for the possibility of raising the retirement age. Thus modified measures of the dependency rate are possible: for example, the sum of the population aged 70 and above, together with those below 20, divided by the population aged 20 to 69.

Even so, the dependency ratio remains a very crude indicator. Aside from adjustment of retirement ages, the tendency for female labour force participation rates to rise over time can greatly modify the effective dependency ratio. The 'real' dependency ratio must also take into account trends in unemployment and productivity; rising productivity per worker could well mean that 'real' dependency falls even as 'demographic' dependency rises.

How are dependency ratios affected by a country's passage into and through the later transitional phase of demographic transition? The answer is fairly straightforward, though sometimes modified to a considerable extent by the factors just mentioned. The typical pattern is that while fertility remains high, the dependency burden is very high. Declining fertility greatly reduces the youth dependency burden without affecting the old age dependency burden very much for some time (see Figure 1, calculated from United Nations 1999). Thus for quite a long period of time the effect on dependency is unambiguously positive. Eventually, the dependency rate bottoms out and finally begins to rise again as the proportion of the elderly begins to increase significantly. The time period in which declining fertility causes overall dependency rates to rise depends on the initial population structure, the speed of the fertility decline and to some extent on trends in mortality.

The long-term trends in dependency ratios for a number of countries, presented in Figure 1, show that the dependency situations reached in 2000 varied widely, depending mainly on when significant declines in fertility began. The ratios in 2000 were approximately:

- Indonesia: .57
- Iran: .77 (and dropping very sharply)
- Pakistan: .83 and barely changing
- Philippines: .70 and declining slowly
- Singapore: .40, having bottomed out and started to rise
- Thailand: .47 and still falling

The 'demographic bonus' in Singapore and Thailand, and to a lesser extent in Indonesia, was very substantial; whereas in Pakistan the burden of a high-fertility age structure was still very apparent. Iran and the Philippines faced much the same burden in 2000. The difference

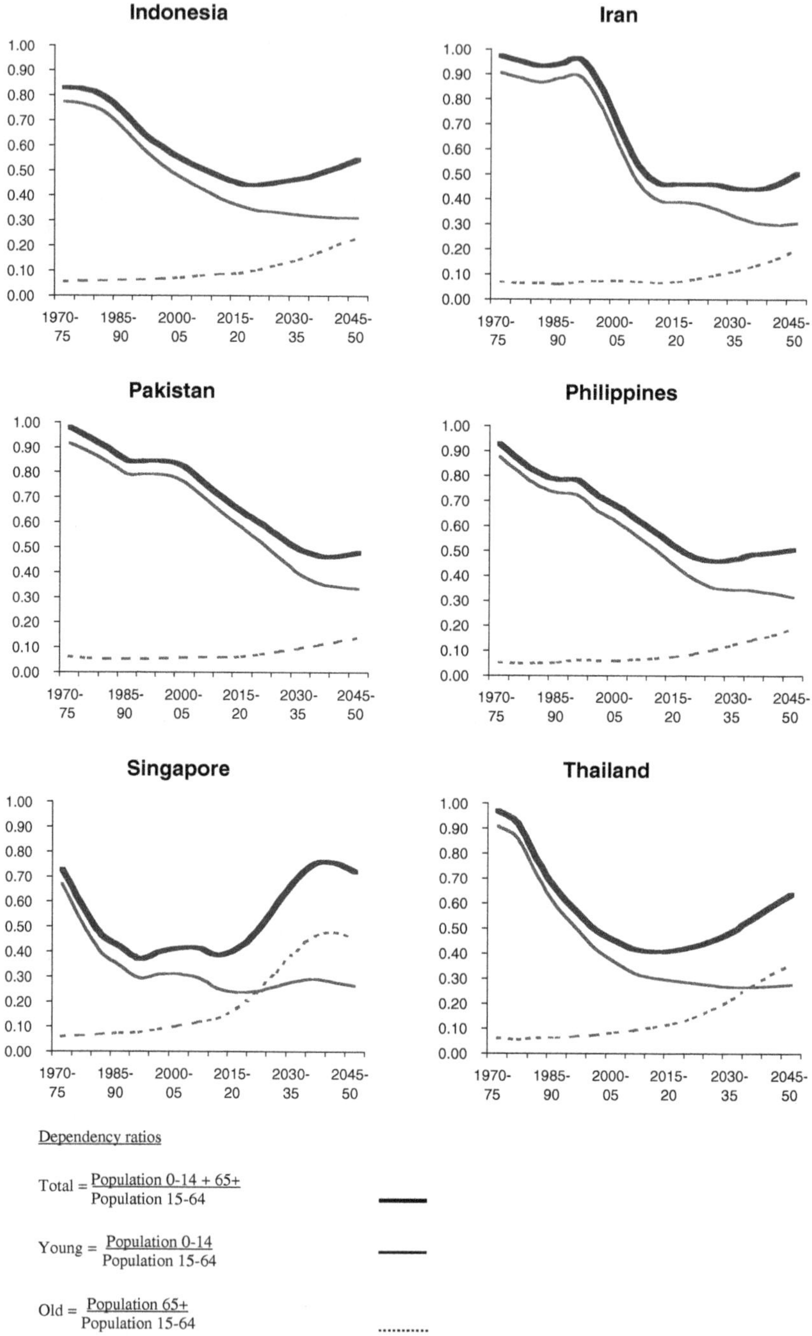

Figure 1. Dependency ratios for Indonesia, Iran, Pakistan, Philippines, Thailand and Singapore (1970–75 to 2045–50).

was that Iran was just beginning a spectacular decline in its dependency burden, whereas in the Philippines sluggish fertility declines were making for only slow amelioration of dependency burdens.

However, there is not much point in having rapid increases in the working-age population if the new labour force entrants continue to have only low levels of education and there are no jobs for them. Pakistan presents a stark example. The labour force is growing very rapidly, and could grow even more rapidly if the female labour force participation rate rose from its current very low levels. The educational level of the workforce is very low, and unemployment rates are high. Average productivity of those employed is also very low.

Educational Development and Job Availability for the Educated?

Countries in the transitional stage are still experiencing rapid increases in their workforce and this is why their dependency ratios are declining. There comes a time in this process when the number of entrants to the labour force ages falls off rapidly – indeed, in some cases begins to decline – but because the workforce covers a broad age range, it takes a decade or so beyond that time for the overall growth of the workforce to slow markedly.

The deceleration and, frequently, cessation of growth of numbers of new labour force entrants is obviously a significant point in the demographic transition. But its implications for employment and educational policy go far beyond the obvious point that the working-age population will never again grow as rapidly as before. Certainly a lightening of the task of finding employment for a rapidly growing workforce is highly desirable. However, before taking comfort from that apparent implication of the demographic trends, four complicating factors need to be considered:

- The overall growth of the working-age population slows much more slowly than do the numbers reaching the working ages;
- Changing labour force participation rates can modify these trends as they translate into the actual growth of the labour force; in particular, the female labour force participation rate (LFPR) is continuing to rise in many countries;
- If the numbers entering the working ages are disaggregated into different educational attainment groups, it is frequently found that the numbers of better-educated entrants are continuing to increase rapidly, even though the entrants as a whole are not increasing;
- The age structure of the working-age population gradually becomes 'top-heavy' as the formerly pyramidal age structure is undercut; a much higher proportion are in the older working ages (45–64) than formerly.

The last two of these points deserve further elaboration. With regard to the educational composition of new labour force entrants, the rapid increases in educational enrolment ratios (facilitated by the decline in fertility) means that new entrants to the workforce with completed high school or tertiary education may continue to increase very rapidly. This raises issues about occupational 'mismatch' and educated unemployment. Simply put, the matrix of occupations by educational level shows a strong concentration of educated workers in the professional, managerial and clerical (P, M and C) occupational groups, which between them employ less than 10 percent of the workforce in countries such as

Thailand and Indonesia, though this rises to above 40 percent in Western countries (Jones 1993). It is unlikely that employment opportunities in such occupations will increase as rapidly as the number of educated young workers entering the labour market.

Logically, if the number of educated labour force entrants is growing much faster than the number of P, M and C occupations, one or more of the following will happen: a higher proportion of these workers than before will enter occupations generally considered 'too low' in status for educated workers (the so-called 'pushdown effect'); and/or, unemployment rates for these educated workers will increase. If the 'pushdown effect' occurs, it will have implications for less educated workers as well, because their 'traditional' jobs are being invaded, as it were, by the better-educated workers.

There are many things that can be said about such a situation. One is that the nature of many occupations changes over time, in the direction of making the occupation more acceptable for workers with more education. For example, workers with a high school education are likely to be more willing to be a sales worker in an air-conditioned department store than in a roadside stall or as a street vendor, or to work as a taxi or bus driver rather than pedalling a trishaw. Another is that expectations are likely to gradually adjust to the realities of the labour market, so that not all high school graduates will expect to get a government job as if by right. Wage differentials by level of education are also likely to narrow if the numbers of better-educated workers are in over-supply, thus sending out signals to deter some young people from continuing to higher levels of education.

Nevertheless, a key policy issue arises: should the government be going all out for educational expansion in such a situation? Clearly, there are many purposes of education aside from the instrumental one of matching workers to job opportunities. On the other hand, however, no government can ignore the possibility of a growing group of unemployed and dissatisfied young people, with potential to cause political unrest. Therefore the issue of what rate of expansion of educational opportunities to plan for cannot be avoided.

A study in one of the poorest provinces of Indonesia, Nusatenggara Timur, highlights the issues (Jones et al. 1998). Young people reaching the final year of senior high school typically hope to find work in government employment. The reality, however, is that fewer than 10 percent of them will be able to find work of this kind. There are very few other job opportunities in large enterprises, simply because there are hardly any large enterprises in this province. The options for these young, well-educated workers are limited. They can go on to higher education, but with the same problem of limited work opportunities after that. They can look for government employment, and if they cannot find it, they can try to find a niche for self-employment, or move to other parts of Indonesia to seek work, or move overseas (especially to Malaysia) to seek work.

One could argue that for this particular province, the Indonesian government should hold back on expansion of the educational system. But this would go against the ideology of providing relatively equal access to educational opportunities throughout the country.

Let us now turn to the final point in the bulleted list above: the increasingly 'top heavy' age structure of the working-age population. Figure 2 (calculated from United Nations 1999)

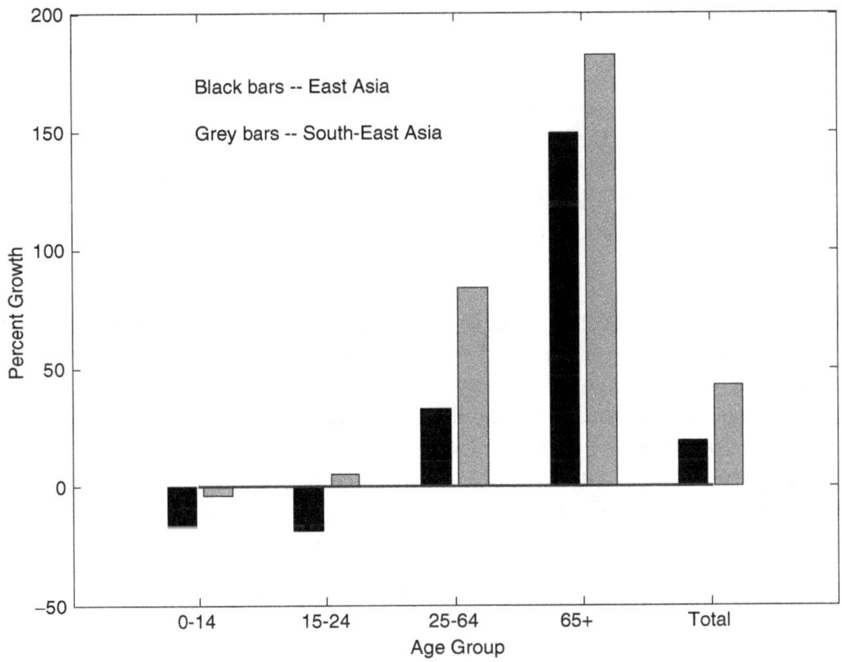

Figure 2. East, South-East Asia: population growth 1995–2025.

shows the projected growth of the broad age groups in East and South-East Asian populations between 1995 and 2025. The sharply higher growth of the older working-age population than of the younger working-age population is the most striking feature. The increasingly 'top heavy' age structure within the labour force has implications for mobility (both occupational and geographic) within the workforce. Older workers are normally considerably less mobile than younger workers. To the extent that a lack of mobility results from lack of education or skills that would enable them to move to other work, attention needs to be given to ways of increasing the job flexibility as well as upgrading the productivity of these older workers who make up an increasing share of the workforce. An obvious need is for more and better designed training programmes – both on-the-job and special programmes – as contrasted to conventional education programmes.

The Transition to Work and Social Justice Issues

One of the most disturbing aspects of the human resource situation in the South-East Asian countries in the later transition phase of demographic transition is the inequality of access to education and to the jobs that education brings within reach. This is of course an issue in every country, but it tends to be more severe in countries where most students do not reach secondary education or, if they do, fail to graduate from secondary education. Those students who miss out tend to be from rural areas and from disadvantaged groups in urban areas. Moreover, the quality of education tends to be lower in schools serving rural and poor urban children.

The disadvantage faced by students from lower socio-economic groups carries through to the job search process for those who do manage to reach higher levels of education. Job search in countries such as Indonesia tends to rely more on connections than on response to advertised job openings and objective testing. This reinforces the advantage the upper socio-economic groups have experienced ever since entering primary school.

The slowing of numbers reaching school age in these later transitional countries should facilitate broadening the coverage of primary and secondary education. This should serve to ameliorate the disadvantage suffered by those in more isolated areas by improving their access to schools. If the greater availability of educational funds per school-age child enables improvements in the quality of education, this should again assist students in the poorest schools, because a relatively small investment in these schools can go a long way towards improving buildings and facilities (though more will be needed to improve the quality of teaching). The example of Indonesia, however, shows that even when the numbers of school-age children do not increase, improvement of access and quality remains a great challenge (Jones and Hagul, 2001).

International Dimensions of the Issue

Where problems of finding employment for rapidly growing workforces are severe, one mechanism sometimes available to lighten the burden is international labour migration. This serves both to lower the threshold of jobs needed domestically to prevent unemployment getting out of hand, and to bring remittances that ease the foreign debt burden. Countries such as the Philippines and, to a lesser extent, Indonesia, have followed this route. However, available jobs for international labour migrants generally require at least some secondary education. Countries such as Pakistan, then, where educational levels remain very low, cannot capitalise fully on these opportunities.

In future, the prospects for absorption of labour internationally will become increasingly positive. The expected decrease in the labour force in some large European countries such as Germany and Italy by about 20 percent between 2000 and 2030, and in Japan by 15 percent over the same period, will put strong pressure on them to recruit workers internationally to meet the labour requirements of their economies. But in general, tertiary education will be required for these jobs, stressing once again the need for educational advance to be made as quickly as possible by countries wishing to make the most of these opportunities.

Fertility Trajectories and Cohort Succession

Many countries passing through the demographic transition experience a fairly smooth decline in fertility. In some cases, however, there have been ‘pauses’ in fertility decline and, in others, even a rise followed by a renewal of decline. Examples of such rises include China from 1962 onwards, following the low fertility of the famine years 1959–61, Iran in the late 1970s to late 1980s, Singapore from 1986 to 1993 and Malays in Malaysia from 1978 to 1986. Such temporary rises in fertility can lead to 'echo effects' in age structure caused by an increase in births when these larger birth cohorts enter the main reproductive ages. Basically, a rebound in fertility from moderate levels is likely to have more impact than a rebound from very low levels, because the rebound from very low levels is not likely to be very large in absolute terms. In cases where the 'echo effects' are substantial, this can have implications for educational planning, labour market trends and other relevant areas.

Even without 'echo effects', cohort succession can lead to very different growth rates of different age groups within the population (Figure 3, calculated from United Nations 1999). For example, in Singapore, the age group 5–14 declined between 1980 and 1990, rose by almost 30 percent between 1990 and 2000, and is expected to decline again over the next two decades. Moreover, while it was increasing sharply over the 1990–2000 period, the age group 15–24 was declining sharply. The contrasts in Iran, both over time for the same age group and at the same time for different age groups, are even more dramatic. These stemmed from the sharp rise and fall in Iranian fertility.

Planners do not always take any notice of age-structure effects, nor do academic economists (e.g. some influential growth models assume the rate of growth of the labour force is the same as the rate of growth of the population). The scale of the shifts shown in Figure 3 are large enough to be taken into account in the planning process (see also Pool, 1994).

In Sweden, there was a fertility peak in the 'baby boom' period, and this is having major, but temporary, effects on the need for teachers. The government wanted to produce more teachers to deal with the upsurge in student numbers. But recruitment of teachers is not so easy. Young people apparently resisted being trained as 'extra' teachers because they were aware that the numbers of school children would soon fall away again, and thought (probably correctly) that their longer-term prospects as teachers would therefore not be very bright.

Singapore

The example of Singapore illustrates the interplay of age structure changes and fertility fluctuations in accentuating or damping down 'echo effects'. The Singapore fertility decline over the 30-year period 1957–1986 was so sharp that it led to significant decline in annual births up to about 1977, despite the population momentum built into the Singapore age pyramid. From 1977 to 1987, the number of births was fairly steady, with continuing slow declines in fertility interacting with a slight decline in numbers entering the key child-bearing age groups. From 1988 to 1994, annual births were up a bit, as a result of a policy-induced slight rise in fertility interacting with steady numbers entering the child-bearing age groups (see Saw Swee Hock 1999: Table 9.1). From 1994 to the present, annual births resumed a slight downward trend. Figure 4 (calculated from Saw See Hock 1999: Tables 9.2 and 9.9; United Nations 1999) shows the trends in births, total fertility rates and numbers of women in the key reproductive ages 20–39.

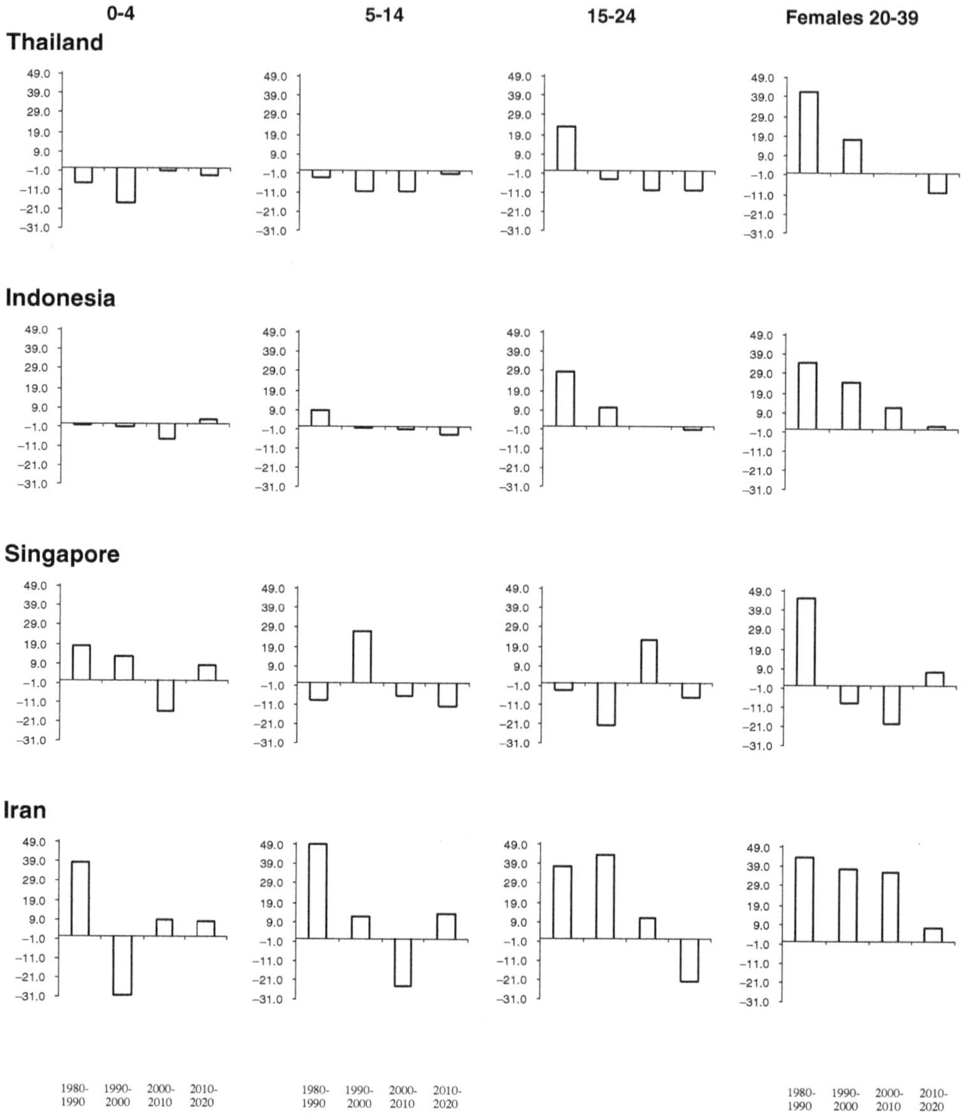

Figure 3. Decadal change in size of selected age groups 1980–2020.

These birth trends necessitated a contraction in maternity hospital facilities, especially over the 1963–75 period, and subsequently primary school facilities over the 1969–1984 period. A contraction of junior secondary school facilities would have been in order, over the 1974–1987 period, except that enrolment rates at this level were still rising, and this helped to modify the downward trend in enrolments. No major 'echo effects' were evident because the rise in fertility after 1987 was not very marked, and also because the trends were modified to some extent by migration.

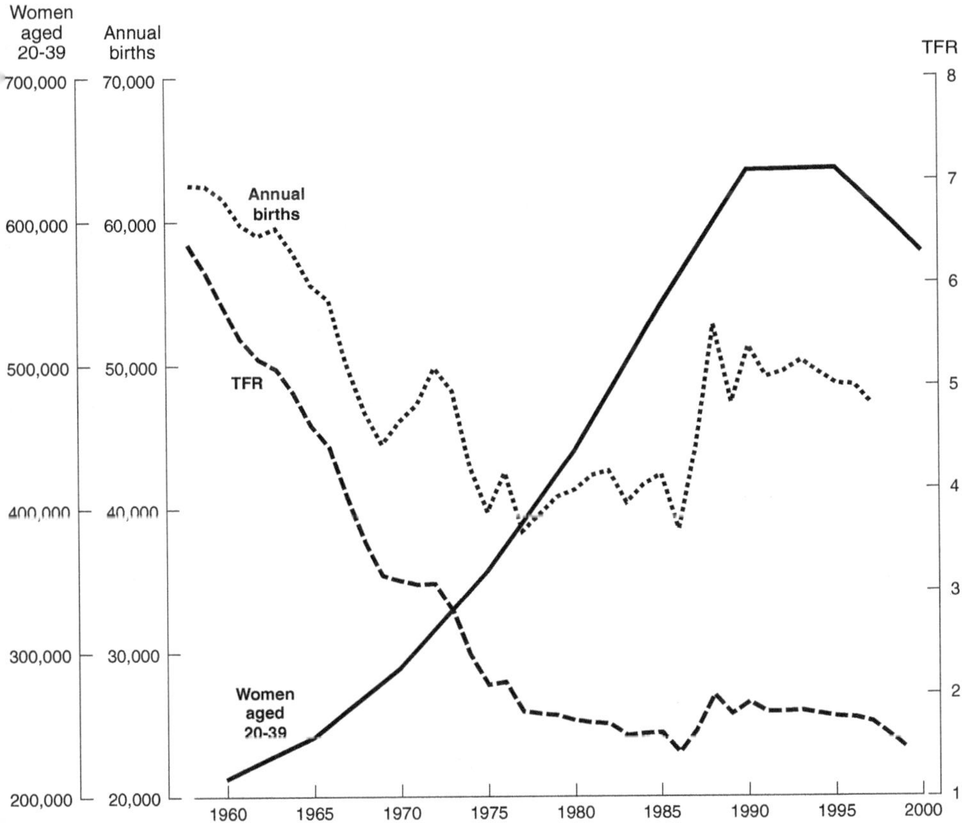

Figure 4. Singapore: Trends in number of women aged 20–39, total fertility rates and number of births, 1958–2000.

Iran

Unlike Singapore, Iran turns out to be a spectacular example of 'echo effects', resulting from a sharp upsurge in fertility, followed by a very rapid decline. This left Iran's age structure with a sharp spike in the 5–14 year age group, and to a lesser extent in the 5–19 year age group, in 1996. We do not yet know exactly how sharp the subsequent undercutting will be: that depends very much on fertility trends at the present time, which are not yet clearly apparent. What is apparent, however, is that within the next few years, the beginnings of an 'echo effect' of rising births is to be expected.

The Iranian demographic situation is shown in Figure 5. The projections on which this figure is based assume that fertility in Iran will decline slowly from a TFR of 2.62 in 1996 to reach replacement (2.08) in 2051 (the projections were kindly supplied by Peter McDonald and Rebecca Kippen, Demography Program, Research School of Social Sciences, Australian National University). Survivorship rates are projected to rise steadily. This is a 'no surprises' projection, in sharp contrast to the unpredicted swings in fertility in the

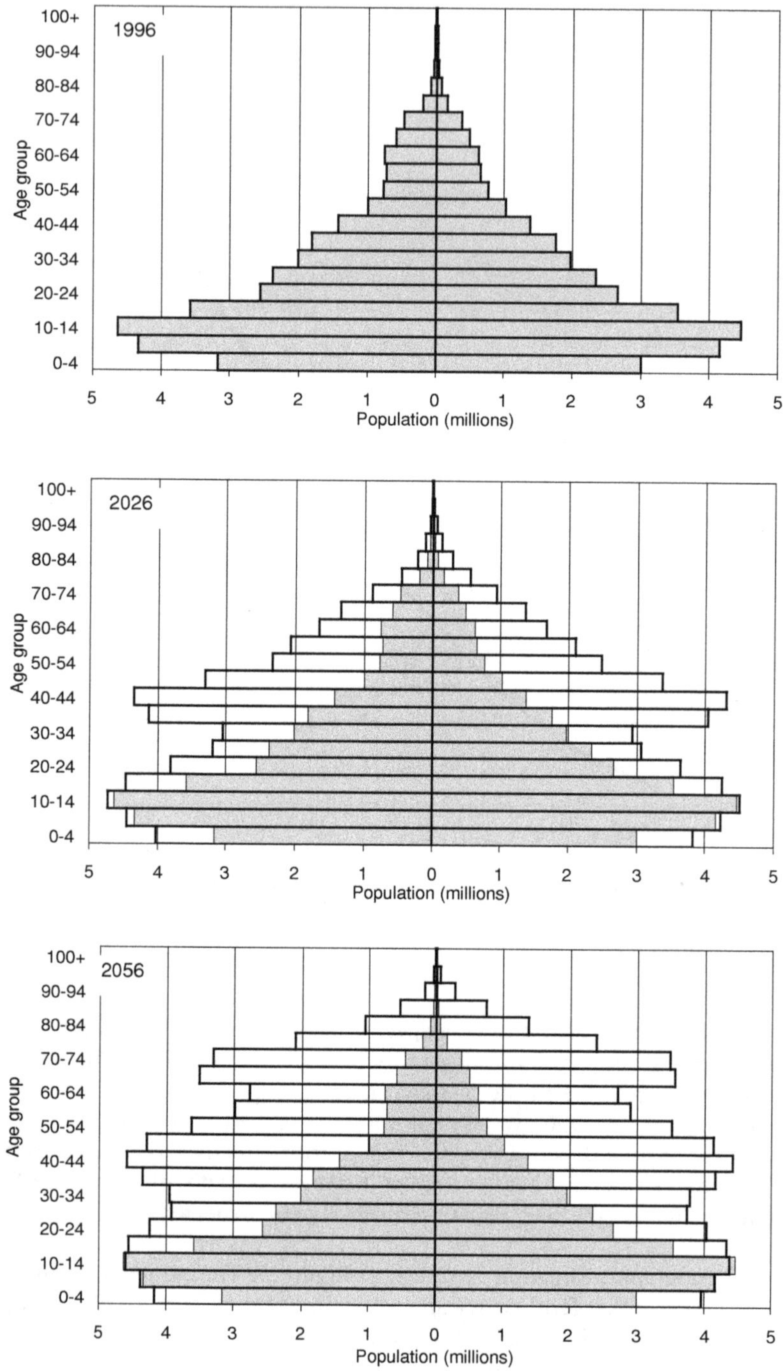

Figure 5. Population pyramids, Iran, 1996, 2026, 2056.

period prior to 1996. The lack of surprises enables the impact of earlier surprises to be traced through the age structure. This impact turns out to be very marked indeed. The spike in the age structure at ages 5–19 in 1996 is echoed again at ages 5–19 in 2026 and, most interestingly, at ages 5–19 in 2056. Thus with a mean length of generation of about 30 years, there is a second and then a third echo 30 and then 60 years after the initial condition.

Reality, in the form of fluctuations in fertility and mortality rates, and international migration, will no doubt intrude to 'soften' these 'echo effects'. In theory, such fluctuations could actually *reinforce* the echo effects, but this would require the coincidence of upward fluctuations in fertility with the passage of the largest group of women through the prime reproductive ages. However, Figure 5 does show that the potential for ongoing echoes is imbedded in the age structure, and that this potential will be realized in a 'surprise free' world.

Conclusion

The human capital aspects of economic development are complex. It can be argued that increasing the quality of human capital is a necessary, but not a sufficient, condition of economic development. The Philippines experience is one where the advantages of a reasonable human capital endowment were frittered away through an inappropriate macroeconomic policy and adverse political factors (Herrin and Pernia 2003). The Philippines has also slid down the 'league table' of human development over the past two decades, with adverse implications for its future economic development.

The two most important lessons of this paper are that: (1) rapid fertility decline typical of the later stages of the demographic transition has positive implications for economic development, among other things through facilitating human resource development; and (2) the age structural changes consequent on the particular fertility and mortality trajectories followed need to be studied carefully and acted on by planners. Cohort succession can follow a quite erratic path where these trajectories have not been smooth, and if ignored, this can lead to a considerable waste of resources.

References

Barro, R.J. (1991), *Economic Growth in a Cross-Section of Countries*, *NBER Working Paper*, No. 3120, National Bureau of Economic Research: Cambridge, MA.

Herrin, A.N. and Pernia, E.M. (2003), "Population, human resources and employment." In: The Philippine Economy: Development, Policies and Challenges, A.M. Balisacan and H. Hill, eds, Oxford University Press: New York.

Higgins, M. and Williamson, J. (1997), "Age structure dynamics in Asia and dependence on foreign capital." *Population and Development Review*, 23(2):261–294.

Jones, G.W. (1976), "The influence of demographic variables on development via their impact on education." In: *Economic Factors in Population Growth*, A.J. Coale, ed. Macmillan: London.

Jones, G.W. (1993), "Dilemmas in expanding education for faster economic growth: Indonesia, Malaysia and Thailand." In: *Human Resources in Development along the Asia-Pacific Rim*, N. Ogawa, G.W. Jones and J.G. Williamson, eds. Oxford University Press: Singapore.

Jones, G.W. (2000), "Demographic trends in East and Southeast Asia and their social consequences." In: *Social Security and Social Development in East and Southeast Asia SPRC Reports and Proceedings*, P. Saunders, ed., No. 143. University of New South Wales: Social Policy Research Centre.

Jones, G.W. and Gingrich, P. (1968), "The effects of differing trends in fertility and of educational advance on the growth, quality and turnover of the labor force." *Demography*, 5(1):226–48.

Jones, G.W. and Hagul, P. (2001), "Schooling in Indonesia: crisis-related and longer-term issues." *Bulletin of Indonesian Economic Studies*, 37(2):207–31.

Jones, G.W., Nagib, L., Sumono and Handayani, T. (1998), "The expansion of high school education in poor regions: the case of East Nusatenggara." *Bulletin of Indonesian Economic Studies*, 34(3):59–84.

Keyfitz, N. (1989), "Putting trained labor power to work: the dilemma of education and employment." *Bulletin of Indonesian Economic Studies*, 25(3):35–55.

Leibenstein, H. (1971), "The impact of population growth on economic welfare – non-traditional elements." In: *National Academy of Sciences, Rapid Population Growth: Consequences and Policy Implications*, Johns Hopkins Press: Baltimore.

Ogawa, N., Jones, G.W. and Williamson, J., eds. (1993), "Introduction." In: *Human Resources in Development along the Asia-Pacific Rim*, Oxford University Press: Singapore.

Pool, I. (1994), *The World's Caring Capacity: Population Compositional Aspects*, The Independent Commission for Population and Quality of Life: Paris.

Saw Swee Hock. (1999), *The Population of Singapore*, Institute of Southeast Asian Studies: Singapore.

Schultz, T.P. (1985), "School expenditures and enrolment, 1960–1989: the effects of income, prices, and population growth." Background paper prepared for the *Working Group on Population Growth and Economic Development*, Committee on Population, National Research Council: Washington, DC.

United Nations, Department of Economic and Social Affairs, Population Division. (1999), *World Population Prospects: The 1998 Revision*, United Nations: New York.

CHAPTER 4. A COMPARATIVE HISTORY OF AGE-STRUCTURE AND SOCIAL TRANSITIONS AMONG ASIAN YOUTH

PETER XENOS
Population and Health Studies, East–West Center

MIDEA KABAMALAN
East–West Center and Department of Sociology, University of Hawaii

Introduction

Underway across Asia today is an interrelated set of demographic and other compositional changes in national populations of youth, the complete ensemble of which we are calling the 'youth transition'. That the social transformations, personal behaviour and demographic profiles of youth are of considerable importance has long been a prominent theme in the social sciences and in very recent years has become *cause célèbre* among policy-makers and international development agencies. Nearly every recent policy commentary relating to youth begins by noting that the current cohort of youth is the largest ever. Not noted, though, is the fact that the global cohort of youth is about to peak according to UN projections, probably never to be as large again. Moreover, the global youth share had already peaked by around 1985. Also not noted is the great diversity that exists among countries and regions. In fact, many countries will experience rising youth numbers for some time to come, while in others youth numbers had already begun to decline two or more decades ago. Many features of contemporary youth demography are not well understood, and the large-scale studies of youth populations that would cast the necessary light remain uncommon.

One relatively unrecognized but crucially important aspect is the transitional and historically novel character of contemporary demographic changes among youth. In this paper and elsewhere we have utilized the label 'youth transition' to describe these changes, stressing the fact that many of them have the character of one-time, irreversible, non-repeatable processes. These will be recognized as essential attributes of the well-known demographic transition conceptual schema that organizes so much demographic knowledge of social change. Elsewhere we have described the primary changes that make up the youth transition as we are defining it, along with certain secondary, derivative changes that seem to be closely associated (Xenos and Kabamalan 1998a, b). This chapter concentrates on the demography of youth transition and draws upon evidence of social transformation only illustratively. To more clearly identify the changes across Asia and their historical uniqueness, contrasts are drawn with the changing demographic composition of youth in some of the European societies during their demographic transitions. We note also that very similar

S. Tuljapurkar et al. (eds.), Population, Resources and Development, 57–88.

changes are occurring among youth in other developing areas, so that the youth transition may be said to be a global phenomenon among contemporary developing societies.

In this discussion the youth age range is taken to be ages 15 through 24, a widely accepted, general purpose, entirely arbitrary definition of youth which is convenient in terms of data sources, particularly census materials. The evidence we have garnered describes composition rather than process, though the latter can be said to be of more fundamental importance. This is entirely a practical matter. Life course data – on the timing of life-events and their configurations in young people's lives – are not widely available, whereas a compositional approach is supported by enough international, comparable data to allow description and disaggregation.

The forces of change seem often to converge on youth, reshaping the demographic and social compositional features of youth into a new and essentially universal pattern. The demographic and social core of the youth transition is rising numbers and proportions of youth in national populations, rising percentages of youth remaining single until well into adulthood, and rising proportions of youth enrolled in school. But a more complete outline of the youth transition would incorporate a variety of other important changes which are important but less regular in pattern across societies (e.g. changing patterns of labour force participation) or systematic but very difficult to measure long-term (e.g. decline in the age at menses). We cannot directly measure many of the abstract dimensions of change that are of keen interest (e.g. degree of disaffection from traditional public mores, decline of guidance and protection within families, changing commitment to the labour market), but we have identified some social-demographic categories of considerable interest, for which pronounced historical changes and cross-societal variations are apparent.

In policy and programming circles there is a broad awareness of these changes, though not always of the whole complex of changes taken as a pattern. The whole package of changes during the youth transition is closely associated in historical terms with demographic transition and its component social and economic transformations and more diffusely linked with rapid and transforming economic change. The causal connections among these historical events are complex. Some elements of the youth transition precede demographic transition or comprise one of its early stages. Other elements are part of the modern social transformation.

In the next section we highlight some connections between youth demography and policy. Then we turn to an analysis of Asian youth data, incorporating comparisons with historical Europe, another region and time of unique social transformation, whenever that is possible.

1. The Policy Focus on Youth

In recent years policy attention has rightfully turned to youth and their special problems, and to the programmatic challenge of reaching youth with useful information and effective programs. This focus on youth by governments, international agencies and NGOs reflects the general recognition that a nation's youth can be a considerable resource for national development but also a significant source of problems. Nowhere is this new attention to youth more visible and welcome than in the arena of reproductive health. The 1994 International Conference on Population and Development's Programme of Action (United Nations 1994; Basu 1997) stressed reproductive health and empowerment, especially for excluded groups such as women and youth. That focus was then reinforced by the events surrounding ICPD+5.

Our analysis focuses on the societies of Asia, a region notable for complex age dynamics set in motion by its 20th century demographic transitions. Asia is commonly treated as one region but it is actually several. It is an assortment of nations and societies possessing great diversity of economic development levels, religions, and cultures. Its political systems range from democratic to authoritarian. There is considerable policy interest in youth among Asian governments and publics, as is clear when we compare, for example, UNICEF's early regional overview (UNICEF 1967) with ESCAP's review at the end of the 1980s (United Nations 1989), and then compare those with ESCAP's most recent overview (ESCAP 1997). These reflect a UN-orchestrated series of initiatives in this area, from the International Youth Year (1985), to the Jakarta Plan of Action on Human Resources Development in the ESCAP Region (1988), to the World Programme of Action for Youth to the Year 2000 and Beyond (1995), to, most recently, the Beijing Statement on Human Resources Development for Youth in Asia and the Pacific (1996). However, one is struck by the range of approaches to social policy including youth policy (MacPherson 1992). While all governments see their youth both as problem and as resource, there are marked differences in emphasis, from the hands-off, laissez faire approach adopted by the Philippines (Raymundo 1989) and colonial Hong Kong (Central Committee on Youth 1988, 1989), to the attempts at heavy-handed paternalism by Singapore (Quah 1981, 1989; Salaff 1988; Heng and Devan 1992) and Indonesia (Shiraishi 1997).

Though marked by obvious differences in history, social system and economic achievement, Asian societies are subject to common forces as well, which engender a degree of similarity. Perhaps the most influential shared experience of these societies is demographic transition and its many significant corollaries (Bongaarts and Watkins 1996; Freedman 1995; Leete and Alam 1993) from personal changes due to the extension of life, to the transformation of family life due to fertility decline. From demographic transition directly flows age dynamics such as youth bulges and deficits and a host of associated changes among youth, some dimensions and consequences of which are examined here.

1.1. THE POLITICAL AND SOCIAL DEMOGRAPHY OF YOUTH

A dynamic youth demography in the form of youth 'bulges' and 'deficits' is a long-standing theme in macro-societal studies of social and political change. Moller (1968) and Weiner (1971) review the literature into the 1960s, while Goldstone (1991) provides a highly regarded recent example at continental scale with his investigation of two century-long waves of state breakdown across Eurasia. He identifies remarkably similar processes in diverse settings – England, France, the Ottoman and Chinese empires – at about the same time and all culminating at mid-17th and 18th centuries. Rapid population growth is identified as an underlying factor in each instance. The Goldstone model is 'demographic/institutional' in character. The overall number of youth is not especially important of itself, but becomes so in interaction with prevailing social and economic institutions. In England, for example, inheritance institutions are said to create a large pool of non-inheriting later-born sons who are easily mobilized politically when they cannot be absorbed by urban labour markets. Goldstone argues that breakdown occurs when a state simultaneously experiences three problems: (1) state financial crisis due to population growth and population pressure; (2) elite divisions due to heightened competition for positions and resources; (3) a high potential for mobilization of popular groups due to rising grievances and predisposing social patterns. Among the predisposing social patterns is "large numbers of youth in the population," particularly when they are concentrated in urban settings (Goldstone 1991: 136ff).

Goldstone draws on long time-series of population estimates for England (Wrigley and Schofield 1981) to show that the proportions of relevant types of youth were highest in England during its periods of greatest political agitation (e.g. the 1630s; cf. Goldstone 1991: 138ff). The connection to youth numbers in Goldstone's argument is intriguing, though it is not developed in much detail and youth demography plays no more than an implicit role in his discussions of France, the Ottoman Empire, and China. Still, he suggests an interesting line of investigation.

Another recent example of macro-societal, comparative analysis is Huntington's widely debated account of the 20th century resurgence of cultural and religious identity as a political and economic force. He observes that relatively large, dynamic youth cohorts form the vanguard of many of these social movements, notably the 20th century Islamic resurgence (Huntington 1996: 109ff, including Figures 5.1–5.3). Huntington joins analysts for the Central Intelligence Agency (CIA 1990; Population and Development Review 1990) in attaching importance to a benchmark 20 percent share of youth (defined as those aged 15–24) as rising to the level of a 'youth bulge', a social force and political fact with which to reckon. In these analyses a youth share below 15 percent constitutes a 'youth deficit', which is said to be equally problematic because it leads to such outcomes as inadequate new entrants to the national labour force. Much of this discussion has only a limited empirical base, however, never more than time series of the youth share of the total population. Others have also focused on the Arab world and its 'youth explosion' (see Cordesman 1998; Maynes 1998).

Raising different questions and employing somewhat different indicators, Macunovich (2000) follows the work of Easterlin (1980; Easterlin and Crimmins 1985) in seeking to show that relative cohort size is universally a key mechanism in fertility decline. She argues that relatively large cohort size operates on fertility by holding down male wages in relatively large cohorts and thus reducing the demand for children within a supply–demand framework of fertility decision-making. Macunovich's Relative Cohort Size (RCS) index is persons aged 15–24 relative to those aged 25–59. Unlike the youth share measure of the youth bulge, the RCS is unaffected by recent fertility decline and the resulting reduced numbers of infants and children to age 15. Macunovich notes that a rise in the RCS follows a decline in infant and child mortality rates, and precedes the onset of fertility decline. Thus, the RCS results from mortality decline and according to Macunovich is a factor in fertility decline, whereas the youth share indicator rises following and as a result of fertility decline. Thus, a rise in the RCS should precede a rise in our youth share measure.

Especially relevant for our analysis is Macunovich's discussion of how relatively large youth cohorts populate institutions and impinge on their operation. She points especially to labour-market crowding, crowding in families, and crowding in education. In each of these institutional sectors, though perhaps most obviously in labour markets, crowding may operate in a segmented fashion, affecting certain social categories with greater force than others. This relates to a later section of this paper in which we explore the changing social composition of youth over time and differences between countries.

Along the same lines, Bloom and Freeman (1986) employ a "generational crowding" framework and Wriggins (1988) a framework in which youth numbers influence politics through effects on the distribution of scarcities, and through the maldistribution of loads or demands relative to the capabilities to meet those demands. Wallimann and Zito (1984) consider cohort size and youthful protest but go beyond the sheer numbers to focus on those "... who are neither entrenched in the family structure nor integrated into the labour market" (p. 69), and on the rise of "quasi-autonomous" peer groups.

These theoretical and practical issues on a global scale are complemented by a few studies of Asian settings in which the changing demography of youth has figured prominently. The number or growth in numbers of youth, or the youth share of the total population, has at times been described as a major or contributing factor in political disturbances (on Indonesia see Keyfitz (1973; 1986); on India see Butler (1990); on South Korea see Fuller and Pitts (1990); on Sri Lanka see Fuller (1995)). While the simplest of these analyses consider only absolute numbers, or relative numbers, or the rate of change in numbers, a few have recognized the importance of social composition or social demography within the youth population. Keyfitz discusses Indonesia's post-war wave of youth in terms of educational planning, urban migration and job creation (Keyfitz 1973; 1986). In her examination of India, Visaria (1986) devotes attention to aspects of social composition including school enrolment and marital status. Xenos (1990a, b) considers the social demography of Asian youth in a comparative, descriptive framework, as does Jones (1997a, b).

Much of the analysis just cited is focused on one society or looks at a relatively confined period of time. Important differences among societies and significant long-run changes are not emphasized. Comparison is further inhibited by variations in the definition of youth. And, too much importance is attached to the total number of youth and not enough to youth social composition. In this paper we draw upon and summarize our research covering 17 Asian societies, every country (except Vietnam and North Korea) with a population of 2.5 million or greater. For each, we assembled comparable data for persons aged 15 through 24 covering the span of four decades from 1950 through 1990. We disaggregated Asian youth cohorts on the basis of available statistical series to produce an effective level of detail while retaining significant time and geographic-comparative dimensions.

We aim to examine rather than ignore the demographic dynamics that underlie cross-sectional, descriptive youth demography. In particular, we explore the transitory versus permanent characteristics of youth demography, thus avoiding some of the misleading conclusions that derive from a static, cross-sectional view (e.g. Braungart and Braungart 1989; 1990). We note important differences between the historical European and other pre-transition settings considered by Goldstone and the societies of 20th century Asia in the midst of demographic transition examined here.

1.2. 'MOMENTUM' AND YOUTH-FOCUSED POPULATION POLICIES

The future of population growth is largely a matter of future rates of childbearing, combined with present day population composition, future mortality being a relatively predictable and minor source of uncertainty. Present and future fertility is often divided for program and policy purposes into wanted and unwanted components, the latter being the immediate target of family planning programs worldwide. The relevance of present-day population composition lies in the future population growth that is built in when that age structure reflects a past history of relatively high fertility. That future growth potential has been called "population momentum" (Keyfitz 1971). It has been shown that with both wanted and unwanted fertility roughly at today's levels, population momentum accounts for much of the anticipated future growth of population (Bongaarts 1994). Across much of Asia, where fertility is now well below the peak levels of the past, population momentum takes on additional significance. Thus, a recent exercise for the Philippines suggests that about two-thirds of the population growth from 1995 through 2020 will be due to momentum, with wanted and unwanted components of fertility accounting for the remaining third (Herrin and Costello 1996).

In concrete terms, momentum stems from a population having relatively large numbers of young persons, poised to mature into their childbearing years and boost childbearing and population growth substantially. This phenomenon can be called the 'youth bulge', though we recognize that what has happened is partly an excess of youth due to improved survivorship, and partly a deficit of infants and children due to recent fertility decline. In terms of every-day observation there certainly is a youth bulge, which can amount to an almost one-third expansion of the age group's share of total population. If these relatively large cohorts survive through the years of childbearing and produce at roughly the current levels of childbearing in terms of both wanted and unwanted children they will have produced about one-third more births than a cohort of conventional relative size.

Based on these observations, certain policy proposals have been put forward. One core notion seems to be that for given wanted and unwanted fertility levels, or for assumed fertility limited to the wanted level, delayed first births and an older mean age at childbearing would reduce growth due to momentum. It is important to distinguish timing and volume aspects of fertility. The proposed delays would reduce the growth rate of population (and the momentum effect) even without reducing the fertility volume for a cohort. Child spacing will reduce the volume as well, due to life-cycle timetable pressures and intervening opportunities. These linkages are examined in Ruzicka (1979) following the analytic framework given by Trussell, Menken, and Coale (1979).

From this kind of analysis it is recommended that the bulge cohorts that are coming along be encouraged to reduce or delay their childbearing. It is suggested that this can be achieved by steps to encourage (A) marrying later; (B) increasing the age at first birth; and

(C) extending interbirth intervals. The benefits of these changes are not limited to reductions of family size, delayed childbearing, and thus reductions in the rate of population growth. They will also *ceteris paribus* improve maternal and child health, another worthy aim of policy. Extending interbirth intervals (approach C) has long been a goal of family planning programs designed to reach married couples. Later marriage (approach A) seems to be occurring spontaneously almost everywhere and is a core element of the general youth transition as defined here. There has been some policy consideration of induced marriage delay, and governments regularly raise their legal ages for marriage, but none of this seems particularly necessary as population policy. Legal changes have been important in only a few societies (on China see Hare-Mustin 1982), though every society has particular social sub-groups which such legislation would surely benefit. The second policy goal, to increase the age at first birth, generally has been viewed as an automatic consequence of marriage delay, but it is recognized now that in a wide range of settings the reality is more complex than this because of pre-marital sexual exposure and more complex forms of union. On Africa, see Cherlin and Riley (1986), Gage-Brandon and Meekers (1993), and Meekers (1993). On Latin America see Morris (1988) and Singh and Wulf (1990). On Asia see Rindfuss and Morgan (1983), Xenos (1990a, b; 1997), Xenos et al. (2000).

1.3. YOUTH AND THE DEMOGRAPHY OF RISK

The proximate stimulus for the recent focus on youth takes the form of some striking statistics. As judged by a wide range of measures there are rising levels of risk-behaviours of many kinds, from substance abuse to premarital and generally unprotected sexual activity (Senderowitz 1995; McCauley and Salter 1995). Youth populations increasingly are linked to well known precursors of risk behaviour such as disturbed family backgrounds, living away from parents or unsupervised urban ward migration. In the area of reproductive health and sexuality the evidence has been especially alarming, particularly when evidence of risky sexual behaviour is combined with significant prevalence of HIV and other sexually transmitted diseases (Cleland and Ferry 1995; Cleland and Way 1994; Dyson 1990).

There is a general belief that young people today are confronted with an unprecedented range of behavioural choices, many of them involving considerable risk, at the same time that youth are receiving less parental guidance and community support and are taking on greater responsibilities such as for earning income. Thus, substance abuse is easier today than in the past, and in some ways is even encouraged by media-driven youth culture. The same surely can be said of sexual activity. In short, youth are at heightened risk because social and economic institutions are guiding and supporting them less, while demanding more. Moreover, the pace of youth-threatening change seems to have accelerated in the last decade or two.

Thus the intense interest recently in understanding how to reach youth with messages and programs, particularly in developing countries where young people and especially unmarried youth have not been a priority audience in the past. The literature on the problems of youth relating to reproductive health and reproductive risk, and on the interventions that can be most effective in reaching youth, is expanding rapidly (National Academy of Sciences 1997; McCauley and Salter 1995; World Health Organization 1993; Center for Communications Programs 1995). Central to strategic thinking in this area is the notion of reproductive risk and the fact that the total youth population can be disaggregated into risk categories (Lightfoot 1997; Bell and Bell 1993; Koontz and Conly 1994). In media and marketing-oriented thinking about reaching youth the term of choice is 'segmentation'. Segmentation can seek to distinguish fairly subtle lifestyle differences (Slater and Flora 1991), including such matters as connections to peer networks that provide most reproductive health information and can be influenced through peer education efforts (Flanagan, Williams and Mahler 1996; Fee and Youssef 1993), or links with health care professionals (Senderowitz 1997), or links with media messages (Flora, Maibach, and Holtgrave 1995; Nare, Katz, and Tolley 1996).

But the information base for much of the required market segmentation is scanty. Age categories go some way toward identifying critical developmental differences (Austin 1995). Marital status certainly distinguishes quite different risk categories among youth. School enrolment marks a group of youth readily at hand for many kinds of information program (Birdthistle and Vince-Whitman 1997; Education Development Center 1996). Working youth are another segment which can be reached by carefully designed programs. Youth who are out of school (Sikes 1996), or not in the labour force, or not connected to either school or work institutions, represent segments which will be especially difficult to reach. Our analysis provides a detailed comparative description of the changing numbers comprising some of these key social categories of youth.

2. The Demography of Youth Transition

This section focuses on the demographic core of the youth transition. Once we have established the important demographic features of the youth transition we can turn to some closely linked social dimensions. In the following sections relevant demographic changes across Asia from 1950 through 1990 are reviewed briefly, along with projections of these indicators through 2025. Basic data are presented in Xenos and Kabamalan (1998b: Appendix A, Table A.1, first two panels), for each of the 17 countries, each sub-region, Asia as a whole (17 countries, including China) and Asia excluding China. Demographic indicators derived from these basic results are given there in Appendix A, Tables A.2 (population growth rates for age-sex specific groups) and A.4 (sex ratios for age-specific groups). An appendix summarizes the methodology.

Considering first all of Asia, we can describe a historical sequence occurring over the last half of the 20th century and the first quarter of the next. Asian youth numbered 239 million in 1950 but had expanded by 2.4 times to 572 million by 1990, a span of 40 years. That underlying annual rate of growth (2.18 percent) was in line with the underlying annual growth of total population (2.03 percent) over the same period. Between 1990 and 2025, according to the projections, the Asian populations considered here will grow by 39 percent to 3,897 million, but the youth population will grow by only 25 percent to 610 million over the same period. Youth's share of total population was 18.9 percent in 1950 and 20.5 percent in 1990, but will decline to 14.9 percent by 2025. The peak youth population in absolute terms will be reached between 2010 and 2020 and decline thereafter. The growth rate of the youth population had diminished somewhat earlier, and with it youth's share of total population. Since China is often shown separately in our analysis, it is worth noting the same results for Asia excluding China. There were 138 million youth in 1950 but this increased by 2.12 percent annually to reach 322 million by 1990. The projection has the youth population of Asia excluding China at 423 million by 2025 reflecting a much reduced youth population growth rate. The youth share of total population averaged close to 19 percent from 1950 through 1990 but would decline to 16 percent by 2025.

But this regional overview conceals the link at the national level between demographic changes among youth and society-wide fertility and mortality declines. The features we wish to highlight are best seen by focusing on one society. The most important changes and their sequence through time are displayed in Figure 1 for the Republic of Korea. The same diagrams for the other countries are given in Appendix E of the source cited. Korea is highlighted here because it illustrates so clearly the historical sequence of youth demography driven by the demographic transition and particularly by the timing of fertility decline. Korea's high birth rate of the 1950s began a dramatic transition downward starting early in the 1960s. We have used as onset dates for sustained fertility decline the estimates suggested by Bongaarts and Watkins (1996), whose criterion was a 10 percent decline in the Total Fertility Rates (TFR) given in United Nations (1994). The death rate had plummeted earlier following the classic contours of demographic transition. According to United Nations estimates (United Nations 1995: 788) the TFR was 6.07 in 1955 when the Net Reproduction Rate (NRR) was 2.28, but the NRR had dropped below the replacement level by 1990, less than thirty years after fertility reduction had begun.

The demographic youth transition among Korean youth followed in lock-step with its overall demographic transition. The growth rate of the youth population rose dramatically as the large pre-transition birth cohorts reached the youth ages under conditions of low mortality. By the time the youth growth rate had peaked by 1975 at 5.5 percent, the youth share of the total Korean population had risen, from its level of around 17 percent in the 1950s and 1960s to its peak at 23 percent in 1980. During the 1970s the youth population expanded rapidly, and it continued to expand though less rapidly during the 1980s. By 1990 the youth population had peaked in absolute terms and began a long decline. The projection for the year 2025 suggests a youth share of only 12.4 percent, a youth population growth rate of –0.6 percent, and a youth population only 75 percent of the number at its peak in 1990.

The sequence of events just described for Korea and outlined in Figure 1 has been or will be experienced with some variation by every society going through the demographic transition, since it is the necessary outcome of the underlying formal demography linking changes in population size and age composition with trends in fertility and mortality (migration for the moment is left out of the picture). The timing of these events for the other countries reflects the timing of their national fertility transitions. A summary of the key dates for each country is shown in Table 1, where the countries are ordered by the dates by which their fertility declines had begun. Fertility decline began in the early-1960s for the vanguard countries: Singapore, Hong Kong, Korea, and Sri Lanka. Japan had of course experienced rapid fertility decline long before, in the 1920s (Kobayashi and Tsoubouchi 1979). The vanguard countries were followed over the next two decades, in chronological order, by the Philippines, Brunei, Taiwan, Malaysia, Thailand, China, Indonesia, India, Myanmar, Bangladesh, Nepal, and Pakistan.

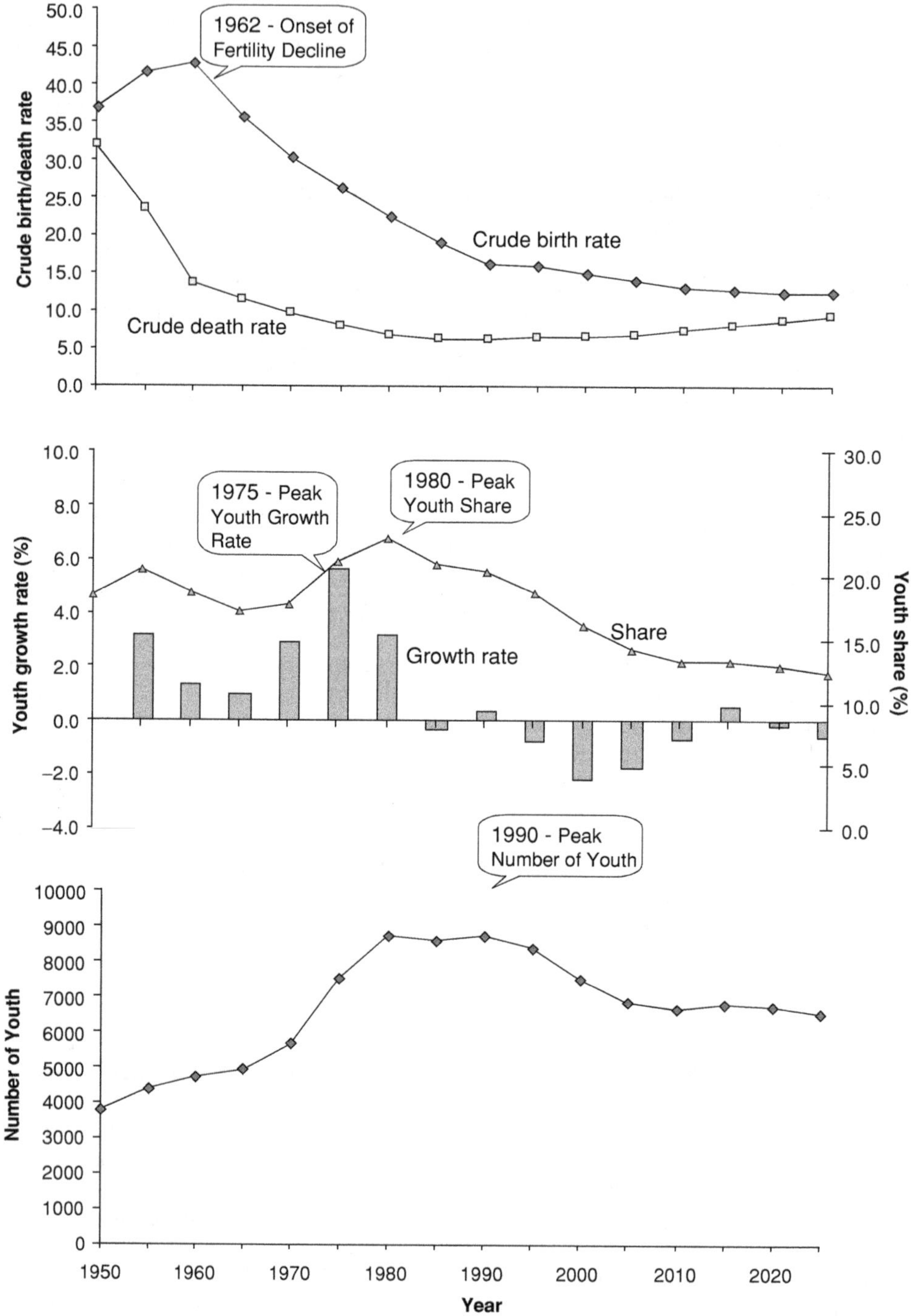

Figure 1. The demographic youth transition: Korea (South).

Table 1. Important dates in the demography of youth transition: countries of Asia.

Country	Onset of fertility decline	Peak growth rate among youth	Peak youth share of total population	Peak number of youth population
Japan	1925	1964	1950	1967
Singapore	1959	1969	1978	1980
Hong Kong	1960	1970	1950	1980
Korea (South)	1962	1974	1980	1981
Sri Lanka	1962	1975	1980	2002
Philippines	1963	1974	1977	2021
Brunei	1965	1970	1980	2012
Taiwan	1965	1960	1980	1980
Malaysia	1966	1970	1980	2015
Thailand	1968	1973	1986	1992
China	1969	1984	1987	1989
Indonesia	1970	1974	1992	2005
India	1973	1977	1984	2014
Myanmar	1976	1985	1994	2001
Bangladesh	1981	1995	2002	2004
Nepal	1988	2001	2007	2032
Pakistan	1990	2005	2010	2033

NB: Onset dates for sustained fertility decline as suggested by Bongaarts and Watkins (1996), whose criterion was a 10 percent decline in the Total Fertility Rates given in United Nation (1994). Their estimates have been supplemented by estimates for Japan, Taiwan, Brunei and Pakistan. Japan's date was taken as 1925, following Kobayashi and Tsoubouchi (1979). Taiwan's was taken as 1965 based on data presented in Chang, Freedman and Sun (1981). Brunei's was taken as 1965 based on data in United Nations (1994). Pakistan's was taken as 1990 based on Total Fertility Rates used in the World Bank projections (Bos et al. 1994).

The peak youth growth rate follows the onset of fertility decline by a decade or so, and the peak youth share then follows by about 20 years. But, there are important variations in the shape of these transitions, and in how distinctly the youth transition shows up in the data. For example, history has provided Japan with a complex demographic transition including a two-phased fertility decline, and its demographic transition is distorted accordingly (Table 1). Hong Kong's youth demography (especially its peak youth share in 1950) reflects its historical periods of very substantial in-migration, a large part of which was comprised of youth. The shifts are greatest, most distinct, and the core demographic features of the youth transition most noticeable when:

- fertility levels were initially high;
- mortality levels had already declined significantly, so there was rapid population growth;
- fertility decline was abrupt and rapid and the decline was to a much lower level; and,
- other factors were not sufficient to disturb the pattern.

Some societies are making the transition with little evidence, while for others the demographic youth transition is distinct, noticeable, and perhaps even a little traumatic in its

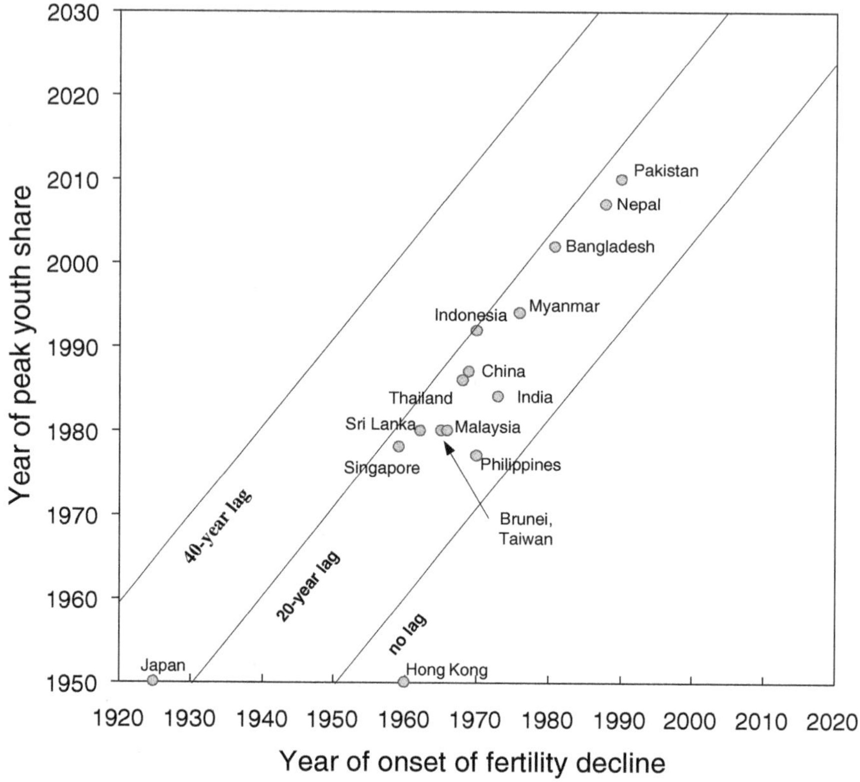

Figure 2. Interval from fertility decline to peak youth share.

impact. Among the demographic features of the youth transition which might stand out are the following:

- a high peak youth share, relative to pre- and post-peak levels, followed by a sharp decline; and,
- a sharp relative increase in the youth population growth rate, followed by a sharp relative decrease in this rate.

Figures 2 and 3 display our 17 Asian countries with respect to the timing of fertility decline and several features of the ensuing youth transitions. This set of figures prompts several observations:

- the peak youth share is experienced about 20 years after the onset of fertility decline; that is, after the numbers of infants and children have been reduced, but before the youth population has been similarly affected (Figure 2);
- for early-decline countries the 1970s were a time of high youth proportions, which peaked in the 1980s (Figure 2);
- for countries with fertility declines beginning in the 1970s and thereafter, peak youth shares have not yet been experienced but are projected for the 1990s or the first decade of the 21st century (Figure 2); and,

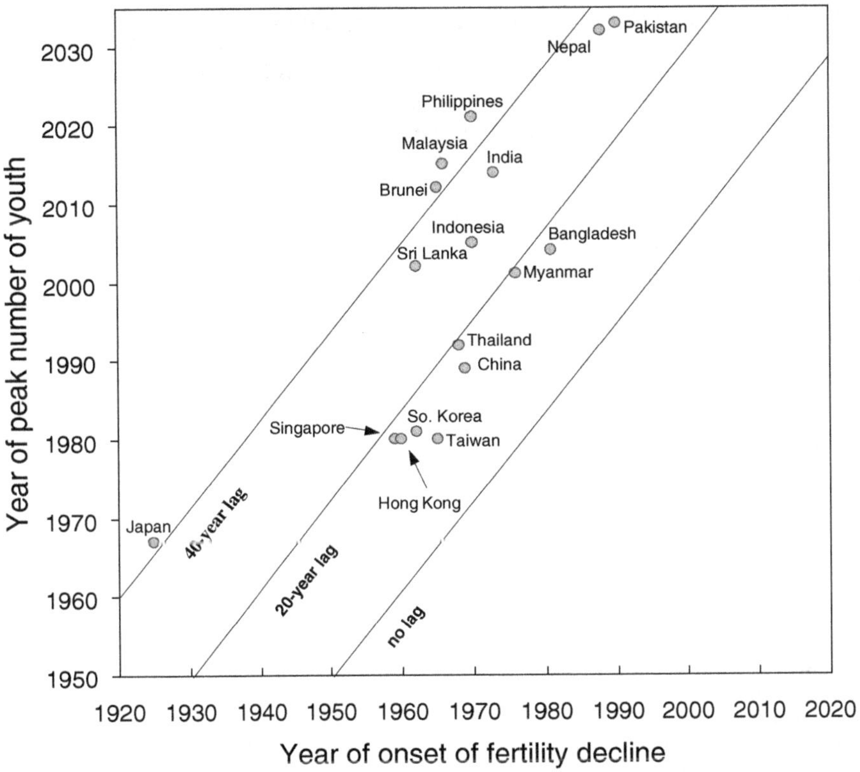

Figure 3. Interval from fertility decline to peak number of youth.

- somewhat later the national youth populations reach their peak absolute numbers and begin to decline (Figure 3). The time lag from onset of fertility decline to onset of decline in youth numbers varies widely (compare Figures 2 and 3). Generally, countries with relatively early fertility transitions have shorter durations from the onset of fertility transition to reaching the peak number of youth. Conversely, countries with relatively late fertility transitions generally take longer to reach their peak numbers of youth.

The more detailed report demonstrates a clear association between the magnitude of the youth share at its peak and the peak growth rate of the youth population; across 17 countries the Pearson correlation coefficient is 0.63. We also illustrate cross-national variation in how precipitously the youth share declines from its peak level, and in the time elapsed between the peak youth share and the peak youth number.

National variations are apparent in Table 2, where countries are arrayed along a dimension involving the details of their fertility transitions and of their demographic youth transitions. While the dates involved certainly cannot be precise, they do convey an essential relationship between fertility transition and demographic youth transition. One group of countries,

Table 2. Patterns of fertility transition and demographic youth transition.

Country	Duration of Fertility Transition[a]	Duration of Demographic Youth Transition[b]	Total Growth of Youth Population During the Demographic Youth Transition (%)
Taiwan	18	15	54.3
Korea (South)	23	18	83.1
Hong Kong	20	20	219.9
Singapore	16	21	111.5
China	21	21	97.2
Indonesia	40	30	103.6
Thailand	32	32	108.8
Sri Lanka	43	38	90.0
Bangladesh	34	39	77.8
India	47	42	106.1
Nepal	42	42	126.6
Pakistan	40	45	100.4
Malaysia	49	44	193.8
Myanmar	49	44	117.0
Brunei	55	60	442.9
Philippines	66	62	258.9

[a]Number of years from NRR = 1.0.
[b]Number of years from onset of fertility decline to decline in number of youth.

comprised most evidently of Taiwan, Korea, Hong Kong, and Singapore, experienced early and rapid fertility declines driven by rapid economic development as well as successful programs of fertility control. This prompted sharply defined, brief, but numerically important demographic youth transitions. In these countries the whole sequence from the onset of fertility decline to the beginning of absolute decline in the youth population took no more than 21 years. China's fertility decline commenced a few years later but it also had a very rapid demographic youth transition. Certain other countries – notably Indonesia and Thailand – have experienced rapid fertility transitions more recently, accomplishing this at somewhat lower levels of economic advance. These countries are also witnessing very distinct and fairly rapid demographic youth transitions, but they have fewer economic resources for social programs, and less rapid rates of absorption of youth into their national economies. A third group of countries are in South Asia – Sri Lanka, Bangladesh, India, Nepal, and Pakistan. They are characterized by late-starting and relatively slow fertility declines taking place at relatively low levels of economic advance. The demographic youth transitions in these countries are much slower and less distinct. Their youth shares are not rising to very high levels, nor are their youth growth rates as high. Myanmar and Malaysia are experiencing even slower demographic youth transitions, and Brunei and the Philippines are in the midst of very slow demographic youth transitions.

Countries with relatively slow, less abrupt transitions, nevertheless produce large proportional expansions of their youth populations. From a policy standpoint, there is a very clear practical trade-off in these results. Especially rapid fertility decline is associated with especially quick and well-demarcated demographic youth transitions. Peak youth numbers and then declines in youth numbers are reached quickly. However, during a rapid demographic youth transition process the youth share and the rate of growth of youth numbers reach relatively high levels. The first situation might be viewed as positive from a social problem standpoint (though not from certain other perspectives – for example, for managing labour markets). The second situation avoids extremes of youth growth and youth shares, but nevertheless allows considerable growth in the youth population before the whole transition process has been completed. The largest numbers in column 3 of Table 2 represent a considerable magnitude of population momentum.

These Asian patterns take on an additional interest when we compare them with the youth age dynamics associated with the European demographic transitions of the 19th and early-20th centuries. We can make this comparison for a number of European countries using census-based age composition data provided in Mitchell's (1982) compilation of historical statistics, together with the estimated timing of European fertility transitions from Watkins (1986). Of the 26 countries examined there, 15 have series spanning the era of fertility decline onset and no major data discontinuities to obscure patterns. In Figures 4 and 5 we present two youth transition indicators for an average of 15 European countries, analogous to the indicators we have presented for Asia. These indicators are the peak youth share, and the growth of the youth population in absolute terms. These are taken to be the peak levels found in the decades immediately preceding or following the onset of fertility decline. All time references have been re-expressed relative to the estimated date of onset of fertility decline, which permits concise comparisons across the two regions (for which regional averages are shown). In Figure 4 we see that the youth share follows a higher trajectory in Asia than Europe with a very evident peak about twenty years after the onset of fertility decline. South Korea is also shown as an example which we will follow throughout this chapter. There the peak is even more apparent.

Figure 5 highlights the much greater growth impact of the youth transition among Asian societies than across Europe. The overall slope for the Asian data is steeper, reflecting the much higher pre- and mid-transition growth rates there than in Europe. The impact of demographic transition on the growth of youth numbers is dramatic relative to Europe. In the twenty years after the onset dates, youth numbers had grown or will grow by 100 percent across Asia, versus about 20 percent across Europe. And, at the fifty-year mark the European youth populations had peaked while the Asian youth populations were still growing. The Asian trend line rises across the whole range of dates shown, but in fact the (projected) pattern shows decline before long – sooner for some countries than others. South Korea illustrates the shape of things when transition occurs early and quickly. South Korea experienced nearly a doubling of its youth population over the 30 years after fertility began to decline in the early 1960s, but has seen a substantial decline in youth numbers since around 1990.

The scatter plot in Figure 6 shows the peak youth share and the peak youth growth rate for each Asian and European country. These are related positively within Asia and across the whole set of countries, but they are not positively related within Europe, and the European levels on each indicator are systematically lower. Only Ireland among the

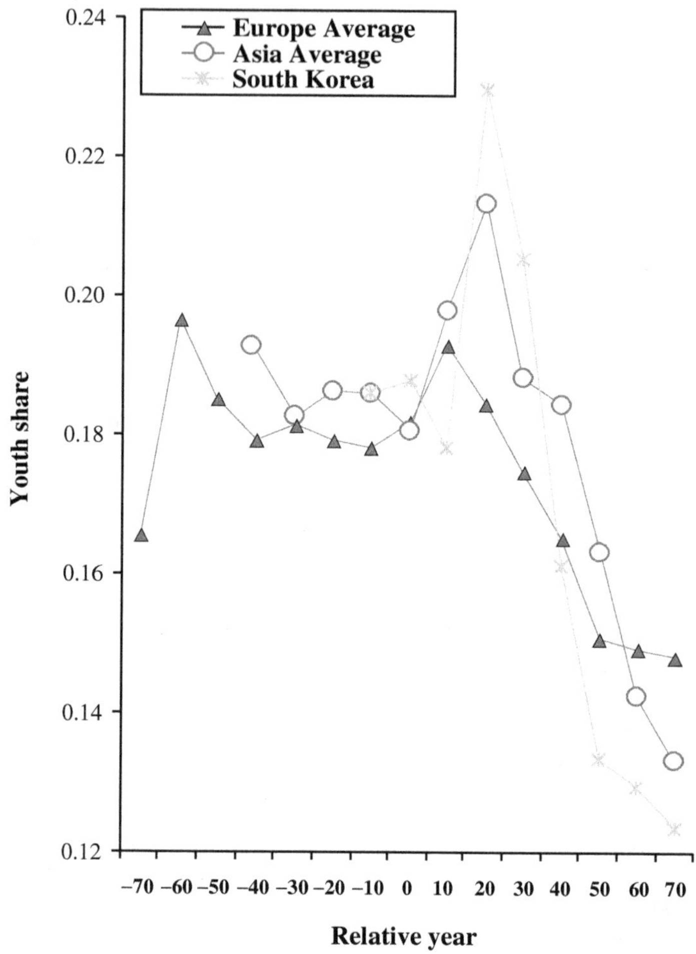

Figure 4. Youth share by relative year: Asian average, European average, and South Korea.

European countries has a peak youth share approaching the Asian experience. Ireland also reminds us that other demographic processes are also at work. Its peak youth growth rate is negative due to out-migration on a large scale. This means that Ireland's out-migration was heavy among youth as well as other age groups, but perhaps greater at somewhat older ages.

The whole demographic transition episode produces considerable population growth overall, though certainly more among Asian than among European countries given the much higher initial fertility levels involved. Youth populations share in this temporary burst of growth, but as Figure 7 shows there is a wide range of experience. Japan's experience was similar to much of Europe. Several Asian countries have had or are projected to have

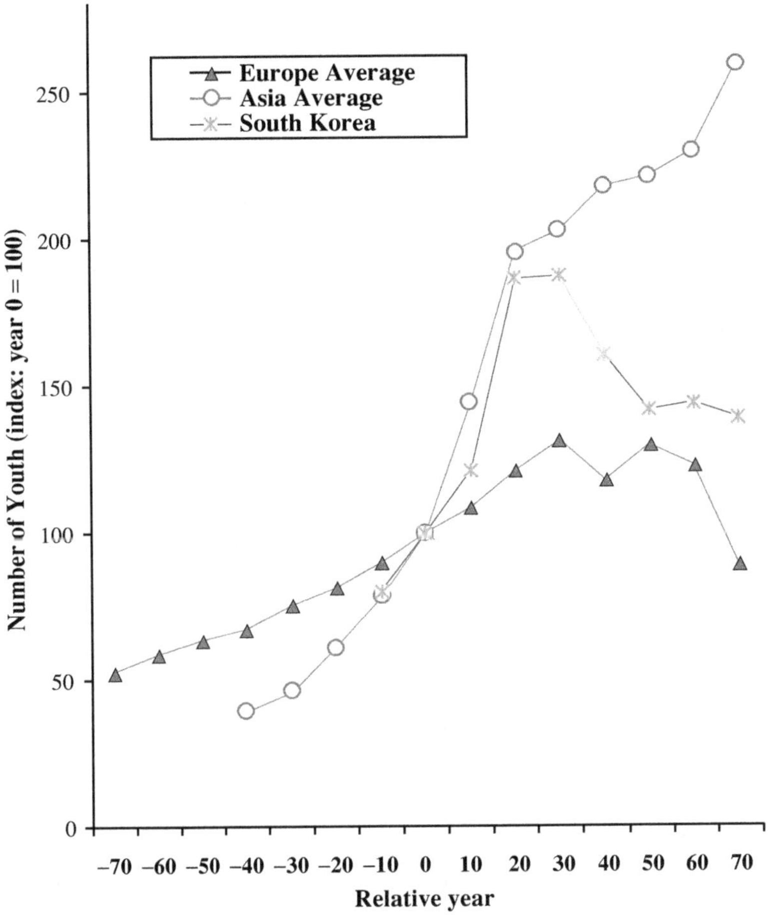

Figure 5. No. of youth by relative year: Asian average, European average, and South Korea.

very large relative increases in youth numbers, and the Asian levels are notably higher on average than the European levels.

These then are the broad contours of the demography of the Asian youth transition. This youth bulge sequence must be seen in the context of the overall transformation of age composition in the course of demographic transition. There are two other, related shifts of age composition which are better known: the decline in proportions in infancy and childhood, and the rise in proportions in old age. Both of these, the diminished stock of children, and the rise of old age, are permanent features of post-transitional society, long-term structural changes, while the youth bulge is a temporary phenomenon. Temporary as it is, though, the youth bulge is difficult to ignore in the short run. This is all the more the case when the changes in numbers of youth are complemented by changes in social

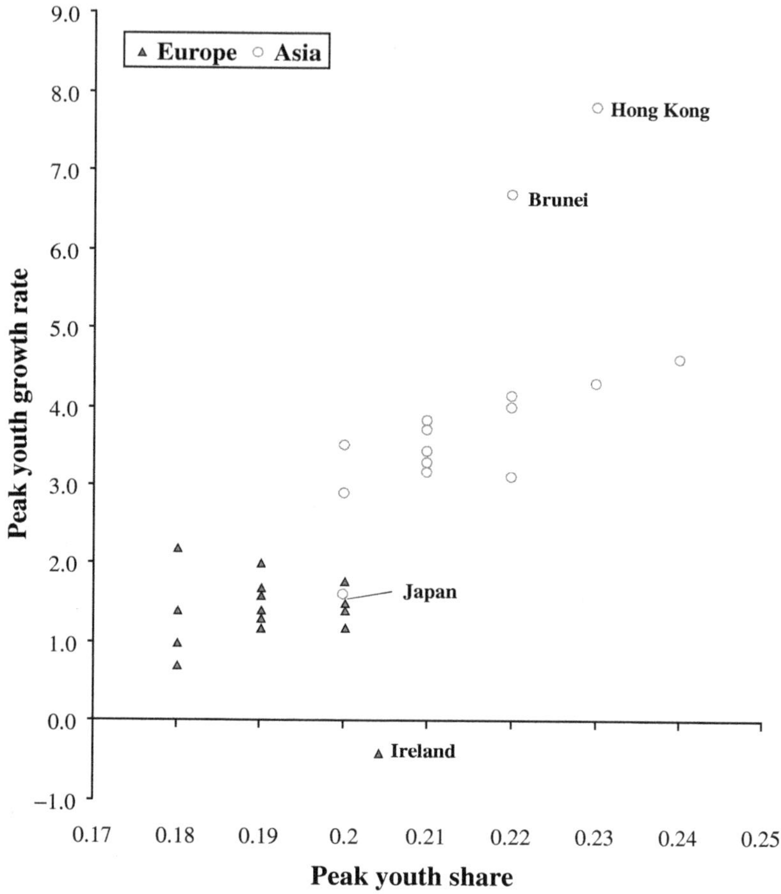

Figure 6. Relationship between peak youth share and peak youth growth rate: Asia and Europe.

composition. Particular sub-groups of youth can expand very rapidly indeed, as we will now see.

3. Modern Youth-Focused Social Transformations

It is beyond our topic to review twentieth century Asian social change comprehensively. However, many of the most dramatic changes are closely associated with two of the three social time series which are available from statistical sources. Together they provide a framework within which to consider many other changes. The youth transition as defined here comprises certain core changes in the social topography of youth. The demographic core of the youth transition was just described. The social change aspects of the youth transition prominently include a shift toward later marriage and rising school enrolments.

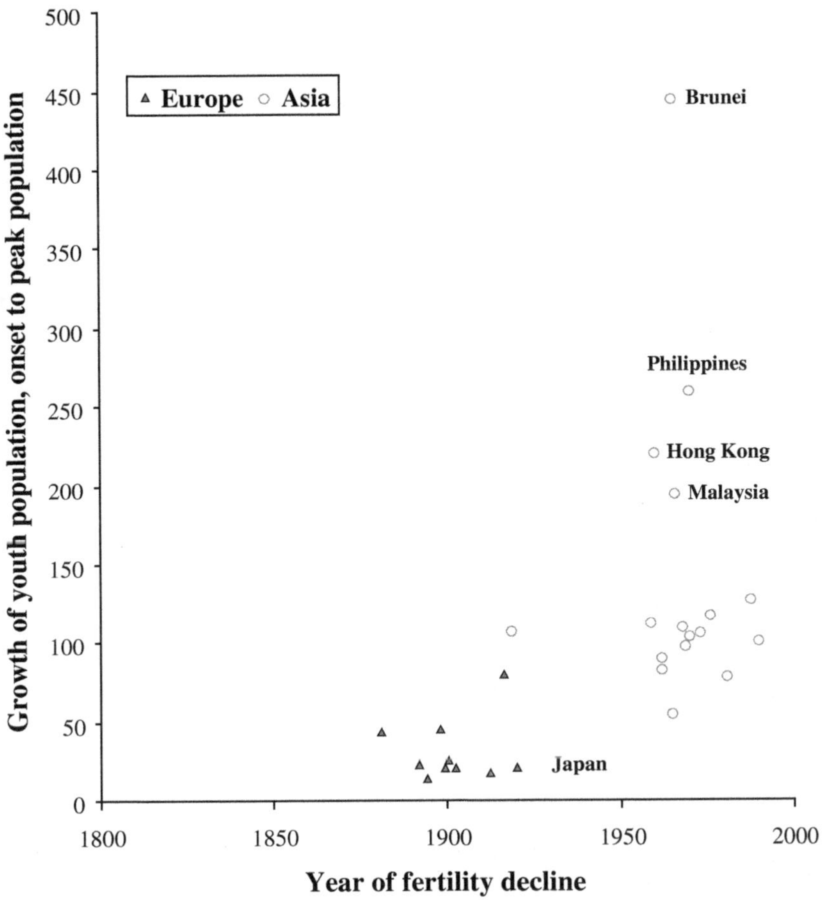

Figure 7. Growth of youth population from onset to peak population by year of fertility decline.

These are both core social transitions, universal, common experiences of one-time, non-repeatable, irreversible transformation among youth across the region.

3.1. THE SHIFT TO LATER MARRIAGE

A considerable body of survey data has been examined for Asian countries showing a historic twentieth-century shift toward later marriage (D. Smith 1980; McCarthy 1982; United Nations 1983; Singh and Samara 1996; Westoff, Blanc and Nyblade 1994). But demographic surveys generally focus on women while ignoring men and can describe only the last two or three decades. Census marital status distributions provide another set of indicators possessing considerable geographic and temporal reach (Smith 1980; United Nations 1988, 1990; Xenos and Gultiano 1992, 2002; Jones 1997b). The full complement of Asian census materials has been examined, expanding and updating the compilation first presented by Xenos and Gultiano (1992).

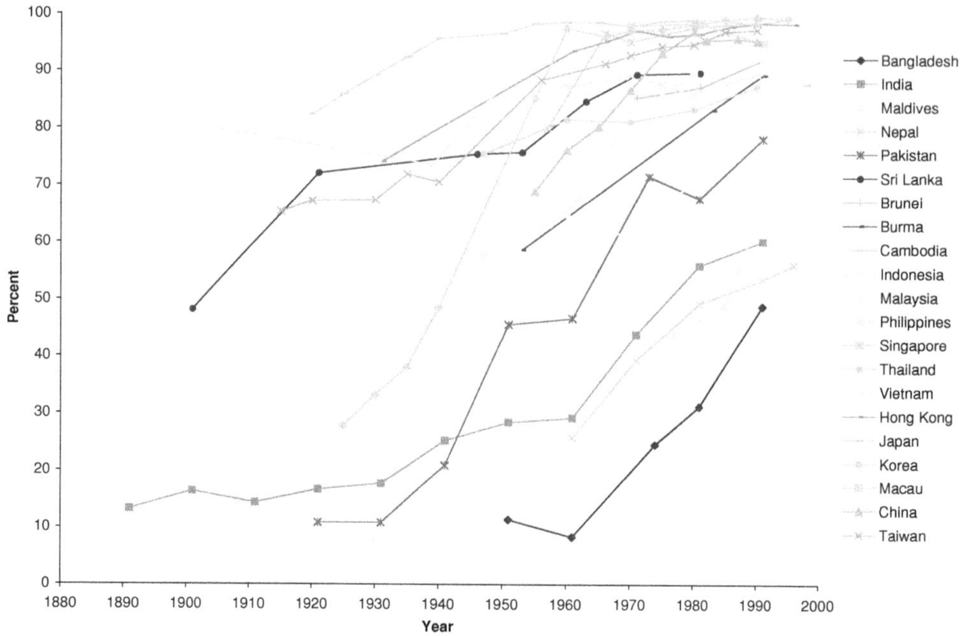

Figure 8. The Asian historical trend toward late marriage: percents single at ages 15–19, females.

The well-documented trend toward later marriage has occurred primarily because percentages of single (never married) people within the youth age range have risen dramatically as illustrated for females aged 15–19 in Figure 8. It is worth emphasizing two aspects of this pattern drawing on the national trends shown in detail elsewhere (Xenos and Gultiano 1992, 2002). First, the pattern is very widespread; it is, in fact, a uniform and universal trend at the national level right across Asia. Second, it has occurred for both sexes, though somewhat more markedly for females than males, leading to some convergence of male and female percents of singles, mean ages at marriage, and the like. Third, even at relatively low levels of GNP per capita, many Asian countries had achieved near-universal marriage delay to beyond age 19. A practical implication of this is that rising percents of singles have occurred relatively early in the development process and therefore are a significant feature of the youth transition, well before and during the period of youth bulge.

These nuptiality changes in every instance supplement the demographic youth transition in adding to the numbers of single youth from year to year throughout the youth transition, until the last stage of the transition when the youth population is declining and the proportions of youth who are single can rise no further. In South Korea, for example, the female youth population expanded by 2.3 times between 1950 and 1990, from 1.87 million to 4.24 million. At the same time the percentages of singles among young women rose dramatically, from 73 to 99 percent among those aged 15–19, and from 16 to 81 percent among those aged 20–24. The result was that the population of young single women in South Korea rose by

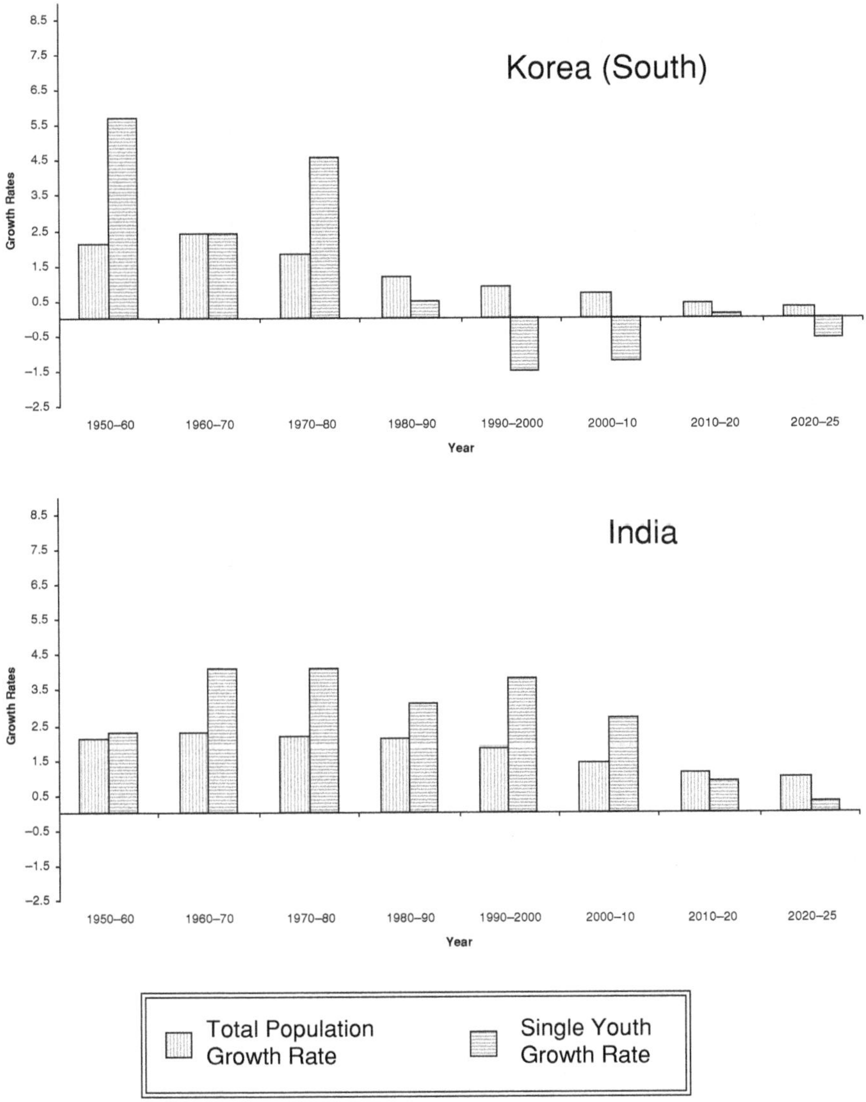

Figure 9. Total population and single youth growth rates.

4.4 times over the 1950–1990 period, from 871 thousand to 3.83 million, at decadal growth rates that at times reached 4–5 percent per year (Figure 9). Not shown but also important is the convergence of female and male nuptiality patterns in South Korea which transformed the relative numbers of single males and single females (Xenos and Kabamalan 1998a, b). In 1950 there were about three single young men for every two single young women. By 1990 that sex ratio was virtually in balance. Similar observations apply to each of the other countries. The data for India in Figure 9 illustrate a similar though less abrupt pattern.

Across all the Asian countries, single population growth rates are sometimes extremely high, both relative to total youth growth rates and in absolute terms (exceeding five or even six percent per year). This is most uniformly so in South Asia (though not in Sri Lanka where marriage delay has occurred throughout the 20th century), but also in Singapore in the 1950s and 1960s and in Brunei in the 1960s and 1970s. For some countries the very rapid growth of single youth is confined to one or two decades, while for other societies moderately rapid expansion of the group was experienced over an extended period of time. A few societies (e.g. the Philippines and Sri Lanka) never experienced extremely rapid growth of the single population because their shifts to later marriage occurred slowly over an extended period.

The essential point is that growth in the number of young singles due to the youth bulge and the growth in the single population due to delayed marriage are reinforcing in their effects. It is important to appreciate the numerical importance of this combination of trends from a programmatic standpoint. For example, between 1950 and 1990 the absolute number of single female youth rose across Asia (excluding China) from 22 million to 82 million. Somewhat more than half of this change was due to the rise in percents singles, without which there would have been 31 million fewer single female youth. Arithmetic of the same kind applies for each country and at more local levels where goal-setting and resource allocation routinely take place.

On this very rapid growth of single populations, a comparison with European societies during their demographic transitions is again helpful as background. Delayed marriage was not a prominent feature of European demographic transitions (Watkins 1986). In fact, marriage was generally earlier for many European countries at the end of the transition than it had been before (Watkins 1986: Table 8.1). Thus, to the much slower European growth rate of youth overall we can add an even slower rate of growth of single youth. We can illustrate this for only one European country, England and Wales, which between 1850 and 1920 saw no important change in female marriage timing. Figure 10 contrasts

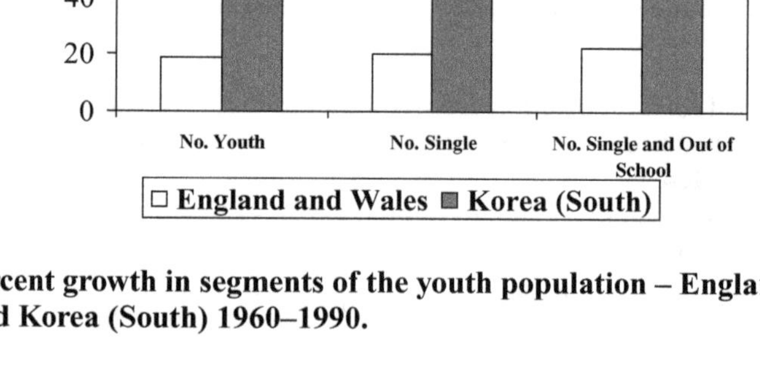

Figure 10. Percent growth in segments of the youth population – England and Wales 1891–1921 and Korea (South) 1960–1990.

the evidence for England and Wales with data for South Korea. Clearly the underlying rate of growth of the youth population is the crucial difference, yet the difference in rates of growth of single populations is even greater. Over thirty years the number of single youth in England and Wales grew by 20 percent, while over the analogous segment of the demographic transition in South Korea the single youth population grew by 112 percent.

3.2. THE RISE IN SCHOOL ENROLMENT

The global, long-term extension of school enrolment from childhood into the adolescent years has affected the youth population of Asia, placing rising proportions in school and diminishing the proportions out of school (Meyer and Hannan 1979; Meyer, Ramirez and Soysal 1992; Benavot et al. 1991; Benavot and Riddle 1988). It is possible to assemble quantitative evidence of this transformation, but not without considerable difficulty. We have compiled or estimated enrolment ratios and percents enrolled for the youth age groups and by sex for the period from 1950 onward through 1990 in 17 Asian countries. Census or survey percents enrolled have been used wherever possible, but it has also been necessary to estimate these percentages from the more widely available enrolment ratios (for the procedures utilized see Xenos and Kabamalan 1998a, b).

The upward trend of school enrolment shares some features with the trend in percents singles. It has been occurring in all the countries, and has been most noticeable among females, resulting in a degree of convergence between male and female percents enrolled. Just as with nuptiality trends, social norms seem to be dictating, at least thus far in the transformation, that male levels be higher. These points can be seen in the national trends in percents enrolled, illustrated here for six countries in Figure 11. The upward movement is greatest in East Asia and least in South Asia, greater for females than males (particularly in Southeast Asia), and of course most notable at ages 15–19 as shown. Though enrolment at ages 20–24 stays low, there is a notable upward trend in East Asia.

The growth rates of enrolled versus total youth populations are presented graphically in Figure 12, again for South Korea and India as illustrations. The full array of data shows that rapid growth of the enrolled youth population, well in excess of the growth of the total youth population, was experienced across East Asia until recently, and that enrolled youth growth rates reached very high levels in much of Southeast Asia over the last two decades or so. Recent policy and long-term trends are both clearly reflected. For example, there is a sharp contrast between the rates for the Philippines, where educational expansion began early, and the rates for Thailand where educational opportunity did not open up to large numbers of youth until the 1960s, or with Myanmar where that opening up has yet to occur. Within South Asia there are important national contrasts. Bangladesh and Nepal had very rapid rates of growth of their enrolled youth populations, while India and Sri Lanka had rates at much lower levels.

Certain observations can be made which parallel those concerning nuptiality. First, the swelling numbers in school supplement the rapidly rising numbers of youth. Staying with the South Korea example, and considering both sexes combined, the number of youth

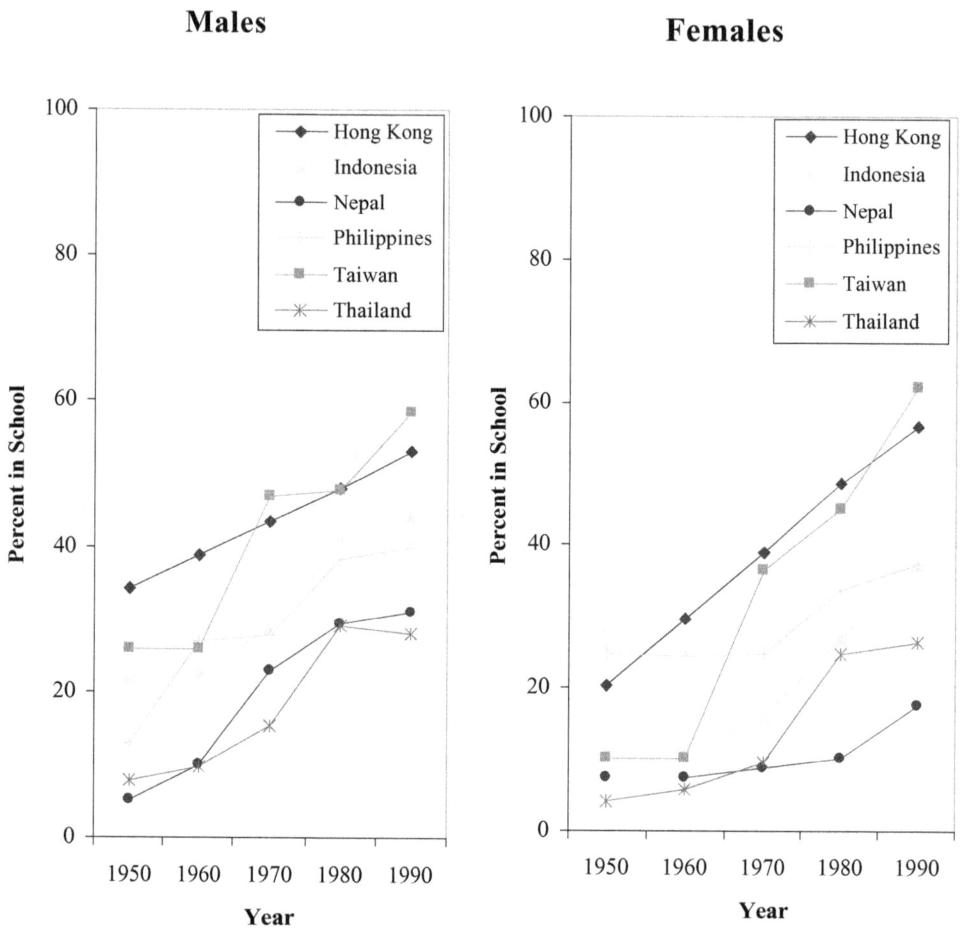

Figure 11. Percent in school at ages 15–19, males and females in six countries, 1950–1990

rose, between 1950 and 1990, from 3.8 million to 8.8 million, an increase of 132 percent. Over the same period the number of youth enrolled in school rose by 600 percent, from 0.6 million to 4.2 million. While that surely placed heavy demands on the educational system, the same changes held down the number of out-of-school youth, which rose only 44 percent from 3.2 million to 4.6 million. This often-problematic segment of the youth population was kept numerically in check because of enrolment trends and despite Korea's very abrupt demographic youth transition. We can also note that growth in numbers unmarried and out of school (a target group for many programs) is much more similar (52 versus 22 percent; see Figure 10) because in South Korea very rapid growth in the young single population is countered by rapid expansion by the number of youth in school.

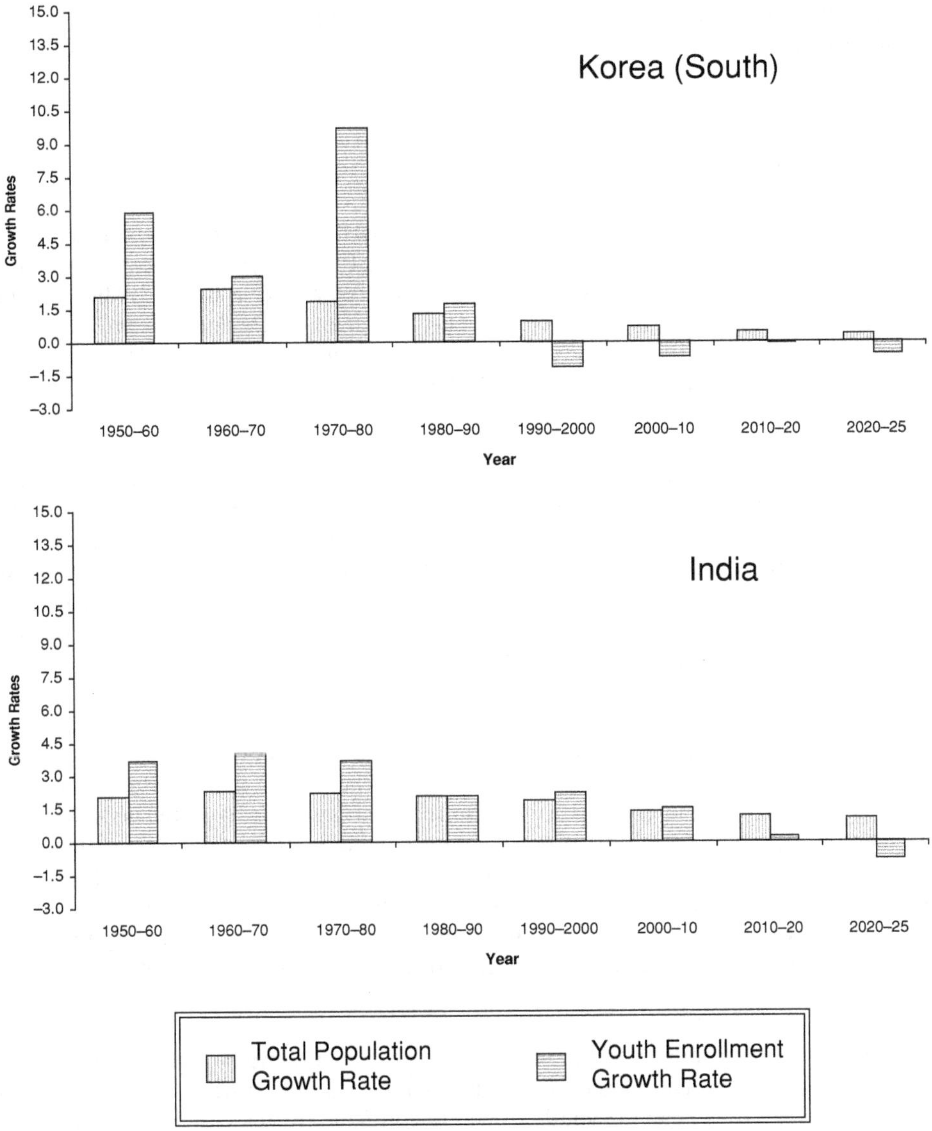

Figure 12. Total population and youth enrollment growth rates.

As one of the vanguard countries with respect to fertility decline, South Korea has already begun to illustrate the dramatic demographic and compositional changes that result from completing the youth transition. Shortly after 1990 South Korea's youth population began to shrink in size and will, according to the projections, decline from 8.8 million in 1990 to only 6.5 million in 2025. Moreover, despite projected increases in enrolment rates (slight, since Korea's enrolment rates were already quite high by 1990), South Korea's in-school youth population is projected to decline from 4.2 million in 1990 to 3.8 million in 2025. The out-of-school youth population will decline even more rapidly, from 4.6 million to only 2.7 million.

4. Summing Up

Much of the epic of social change across Asia has focused on youth, giving rise to a myriad of problems, and there is intense current interest in ameliorating youth policies, especially in the relatively unexplored areas of risk-taking and reproductive health. Policymaking directed to youth issues requires the fullest understanding of the changes that are occurring to youth, including some dramatic changes in numbers and composition. These changes are seen best from a comparative perspective over a historically meaningful period of time.

There is an important element in common across all these youth bulge episodes. In neither late-20th century Asia nor any of the earlier situations examined by Goldstone, Huntington and others is sheer "engulfment by numbers" (Coleman 1974) or the burgeoning absolute count of youth the crucial factor at work. In 17th century England it was young, urban men, especially the later-born men of farm family background who found themselves disenfranchised by the prevailing system of land inheritance. Similar arguments are made for France in the run-up to 1789.

But our interest here is not just in youth as a political force, but also as a diverse audience for programmatic efforts. We stressed at the outset the importance of disaggregating or segmenting that audience so that efforts can be tailored carefully and targeted accurately. Sheer numbers are not the entire story. In the modern-day Asian societies that we have examined demographic transition is producing a notable youth bulge only in those few instances where demographic transition has been extremely rapid and fertility has fallen from a high initial level. The early and very rapid transitions of South Korea, Singapore and Taiwan illustrate this. More important numerically, and much more nearly universal are the systematic and often dramatic shifts in social composition – the youth transition – that has occurred simultaneously.

The remarkable transitions in schooling and marriage that combine with youth demography are more waves than bulges, waves that build on one another to create a crest of remarkable force. At its peak the youth growth rate in South Korea was 5.5 percent. From 1960 to 1980 Korea's youth population expanded by 86 percent, but during the same period of time its single population doubled and its in-school population expanded by 3.5 times.

We have stressed commonalities across the experiences of Asian societies. But unique national experiences are very important as well. The timing, tempo, and force of the transitions vary across countries, and extraneous historical events sometimes obscure the underlying transitional patterns (notably in Japan and Hong Kong). But the shared experience is powerful enough to give the framework some predictive value. Our projections of youth bulges and other aspects of demographic youth transition for South Asia are likely to be born out since they follow directly from demographic transition itself as well as the on-going marriage and educational transitions.

Acknowledgements

The authors acknowledge the able research assistance of He Shi Jen, Mely Silverio and Jin Young Byon. The empirical analysis for Asia is reported more fully in Xenos and Kabamalan (1998b), a report to the Population Council on contract CI97.63A.

References

Austin, E.W. (1995), "Reaching young audiences: developmental considerations in designing health messages." In: *Designing Health Messages: Approaches from Communication Theory and Public Health Practice*, E. Maibach and R.L. Parrott, eds. Sage: Thousand Oaks, CA.

Basu, A.M. (1997), "The new international population movement: a framework for a constructive critique." *Health Transition Review*, 7(4):7–32.

Bell, N.J. and Bell, R.W. (1993), *Adolescent Risk Taking*, Sage: Newbury Park.

Benavot, A., Cha, Y-K., Kamens, D., Meyer, J.W. and Wong, S-Y. (1991), "Knowledge for the masses: world models and national curricula, 1920–1986." *American Sociological Review*, 56(Feb.):85–100.

Benavot, A. and Riddle, P. (1988), "The expansion of primary education, 1870–1940: trends and issues." *Sociology of Education*, 61(July):191–210.

Birdthistle, I. and Vince-Whitman, C. (1997), *Reproductive Health Programs for Youth Adults: School-Based Programs*, Focus on Youth Adults Research Series: Washington, DC.

Bloom, D.E. and Freeman, R.H. (1986), "Youth problem: age or generational crowding?" *Center for Population Studies Discussion Paper*, No. 83.

Bongaarts, J. (1994), "Population policy options in the developing world." *Science*, 263:771–76.

Bongaarts, J. and Watkins, S.C. (1996), "Social interactions and contemporary fertility transitions." *Population and Development Review*, 22(4):639–82.

Bos, E., Vu, M.T., Massiah, E. and Bulatao, R.A. (1994), *World population projections: estimates and projections with related demographic statistics, 1994–1995*. Johns Hopkins University Press: Baltimore.

Braungart, R.G. and Braungart, M.M. (1989), "Youth status and national development: a global assessment in the 1980s." *Journal of Youth and Adolescence,* 18(2):107–130.

Braungart, R.G. and Braungart, M.M. (1990), "Youth movements in the 1980s: a global perspective." *International Sociology*, 5(2):157–181.

Butler, J.A. (1990), "Spatial patterns of age structure and political instability in India." *M.A. Thesis*, University of Hawaii.

Central Committee on Youth. (1988), *Report on Youth Policy*, Central Committee on Youth: Hong Kong.

Central Committee on Youth. (1989), *Report on the Need for a Youth Policy in Hong Kong*, Central Committee on Youth: Hong Kong.

Center for Communications Programs. (1995), "Reaching youth worldwide: a decade of reproductive health programs and activities for young people." *Working Paper*, No. 2. CCP: Baltimore.

Central Intelligence Agency. (1990), "Youth deficits: an emerging population problem." *A Research Paper*, RTT 90-1001 2U. CIA: Washington, DC.

Chang, M-C., Freedman, R. and Sun, T-H. (1981), "Trends in fertility, family size preferences, and family planning practice: Taiwan, 1961–1980." *Studies in Family Planning*, 12(5):211–228.

Cherlin, A. and Riley, N. (1986), "Adolescent fertility: an emerging issue in sub-Saharan Africa." *Population, Health and Nutrition Department Technical Note*, 86-23. World Bank: Washington, DC.

Cleland, J. and Ferry, B. (1995), *Sexual Behavior and AIDS in the Developing World*, Taylor and Francis: London.

Cleland, J. and Way, P., eds. (1994), *Health Transition Review, AIDS Impact and Prevention in the Developing World: Demographic and Social Science Perspectives*, Suppl. to Vol. 4. Health Transition Centre: Canberra. (National Centre for Epidemiology and Population Health in conjunction with the International Union for the Scientific Study of Population. ISSN 1036-4005.)

Coleman, J. (1974), *Youth: The Transition to Adulthood*, University of Chicago Press: Chicago, IL.

Cordesman, A.H. (1998), *Demographics and the Coming Youth Explosion in the Gulf*, Center for Strategic and International Studies: Washington, DC.

Dyson, T. (1990), *Sexual Behaviour and Networking: Anthropological and Socio-Cultural Studies on the Transmission of HIV*, Ordina: Liege.

Easterlin, R.A. (1980), *Birth and Fortune: the Impact of Numbers on Personal Welfare*, Basic Books: New York.

Easterlin, R.A. and Crimmins, E. (1985), *The Fertility Revolution: a Supply-Demand Analysis*, University of Chicago Press: Chicago, IL.

Economic and Social Commission for Asia and the Pacific. (1997), *Review of the Youth Situation, Policies and Progammes in Asia and the Pacific*, ESCAP: Bangkok.

Education Development Center. (1996), *School-based Reproductive Health Programs for Young Adults*, Pathfinder International: Newton, MA.

Fee, N. and Youssef, Y. (1993), *Young People, AIDS and STD Prevention: Experience of Peer Approaches in Developing Countries*, Global Programme on AIDS: Geneva. (draft)

Flanagan, D., Williams, C. and Mahler, H. (1996), *Peer Education in Projects Supported by AIDSCAP: A Study of 21 Projects in Africa, Asia and Latin America*, AIDSCAP: San Francisco.

Flora, J.A., Maibach, E.W. and Holtgrave, D. (1995), "Communication campaigns for HIV prevention: using mass media in the next decade." In: National Academy of Sciences. *Assessing the Social and Behavioral Science Base for HIV/AIDS Prevention and Intervention*, National Academy Press: Washington, DC. (Workshop Summary. Background Papers.)

Freedman, R. (1995), "Asia's recent fertility decline and prospects for future demographic change." *Asia-Pacific Population Research Reports*, 1:1–27.

Fuller, G. (1995), "The demographic background to ethnic conflict: a geographic overview." Edited by Central Intelligence Agency. pp 151–54. *The Challenge of Ethnic Conflict to National and International Order in the 1990's: Geographic Perspectives*, RTT 95-10039. Central Intelligence Agency: Washington.

Fuller, G. and Pitts, F.R. (1990), "Youth cohorts and political unrest in South Korea." *Political Geography Quarterly*, 9(1):9–22.

Gage-Brandon, A. and Meekers, D. (1993), "Sex, contraception, and childbearing before marriage in sub-Saharan Africa." *International Family Planning Perspectives*, 19(1):14–18, 33.

Goldstone, J.A. (1991), *Revolution and Rebellion in the Early Modern World*, University of California Press: Berkeley.

Hare-Mustin, R.T. (1982), "China's marriage law: a model for family responsibilities and relationships." *Family Process*, 21(4):477–81.

Heng, G. and Devan, J. (1992), "State fatherhood: the politics of nationalism, sexuality, and race in Singapore." In: *Nationalisms and Sexualities*, A. Parker, M. Russo, D. Sommer and P. Yaeger, eds. Chapman and Hall Inc: Routledge.

Herrin, A.N. and Costello, M.P. (1996), *Sources of Future Population Growth in the Philippines and Implications for Public Policy*, The Population Council: New York.

Huntington, S.P. (1996), *The Clash of Civilizations and the Remaking of World Order*, Touchstone Books: London.

Jones, G.W. (1997a), "Changing age structure, educational progress, employment patterns, the environment, mobility and their implications for adolescents." Prepared for *Expert Group Meeting on Adolescents: Implications of Population Trends, Environment, and Development, 30 Sept. to 2 Oct., Bangkok, Thailand.*

Jones, G.W. (1997b), "Population dynamics and their impact on adolescents in the ESCAP region." *Asia-Pacific Population Journal*, 12(3):3–29.

Keyfitz, N. (1971), "On the momentum of population growth." *Demography*, 8(1):71–80.

Keyfitz, N. (1973), "The youth cohort revisited." In: *Population, Politics, and the Future of Southern Asia*, F. Guyot, ed. Columbia University Press: New York.

Keyfitz, N. (1986), "The youth cohort and Indonesian history." *Masyarakat Indonesia*, 13(1):1–20.

Kobayashi, K. and Tsubouchi, Y. (1979), "Recent trends and regional variations in Japanese marital fertility." In: . *Fertility Transition of the East Asian Populations*, L-J. Cho and K. Kobayashi, eds. The University Press of Hawaii: Honolulu.

Koontz, S.L. and Conly, S.R. (1994), "Youth at risk: meeting the sexual health needs of adolescents." *Population Policy Information Kit*, No. 9. Population Action International: Washington, DC.

Leete, R. and Alam, I. (1993), *The Revolution in Asian Fertility: Dimensions, Causes, and Implications*, Clarendon Press: Oxford.

Lightfoot, C. (1997), *The Culture of Adolescent Risk-Taking*, The Guilford Press: New York.

MacPherson, S. (1992), "Social policy and economic change in the Asia Pacific region." *Social Policy and Administration*, 26(1):55–61.

Macunovich, D.J. (2000), "Relative cohort size: source of a unifying theory of global fertility transition?" *Population and Development Review*, 26(2):235–61.

Maynes, C.W. (1998), "The middle east in the twenty-first century." *The Middle East Journal*, 52(1):9–16.

McCarthy, J. (1982), "Differentials in age at first marriage." *World Fertility Survey Comparative Studies—Cross National Summaries*, No. 19. International Statistical Institute: Voorburn.

McCauley, A.P. and Salter, C. (1995), "Meeting the needs of young adults." *Population Reports*, 23(3):1–44.

Meekers, D. (1993), "Sexual initiation and premarital childbearing in sub-Saharan Africa." *DHS Working Paper*, No. 5. Macro International: Columbia, Maryland.

Meyer, J.W. and Hannan, M.T. (1979), *National Development and the World System: Educational, Economic and Political Change, 1950–1970*, University of Chicago Press: Chicago, IL.

Meyer, J.W., Ramirez, F.O. and Yasemin, N.S. (1992), "World expansion of mass education, 1870–1980." *Sociology of Education*, 65(2):128–149.

Mitchell, B.R. (1982), *International Historical Statistics: 1750–1970*, Macmillan: London.

Moller, H. (1968), "Youth as a force in the modern world." *Comparative Studies in Society and History*, 10:238–60.

Morris, L. (1988), "Young adults in Latin America and the Caribbean: their sexual experience and contraceptive use." *International Family Planning Perspectives*, 14(4):153–158.

Nare, C., Katz, K. and Tolley, E. (1996), *Measuring Access to Family Planning Education and Services for youth Adults in Dakar, Senegal*, Family Health International: Washington, DC.

National Academy of Sciences. (1971), *Rapid Population Growth: Consequences and Policy Implications*, Johns Hopkins University Press: Baltimore.

National Academy of Sciences. (1997), *Reproductive Health in Developing Countries*, National Academy of Sciences: Washington, DC.

Population and Development Review. (1990), "The CIA on youth deficits." *Population and Development Review*, 16(4):801–7.

Quah, S.R. (1981), "Impact of policy on the family: Can the family be strengthened by legislation?" *Southeast Asian Journal of Social Science*, 9(1–2):33–53.

Quah, S.R. (1989), "The social significance of marriage and parenthood in Singapore." Paper presented at the *XXIVth International CFR Seminar, Singapore, 2–4 May.*

Raymundo, C.M. (1989), "The Filipino adolescents: their implication for Philippine development." Paper written as part of the U.P. Center for Integrative Development Studies—UNFPA Project on Interrelationship of Population, Human Resources Development and the Philippine Culture.

Rindfuss, R.R. and Morgan, P. (1983), "Marriage, sex and the first birth interval: the quiet revolution in Asia." *Population and Development Review*, 9(2):259–278.

Ruzicka, L.T. (1979), Nuptiality and Fertility, Ordina Editions: Liege.

Salaff, J.W. (1988), *State and Family in Singapore: Structuring in a Developing Society*, Cornell University Press: Ithaca.

Senderowitz, J. (1995), *Adolescent Health: Reassessing the Passage to Adulthood*, World Bank: Washington, DC.

Senderowitz, J. (1997), *Health Facility Programs on Reproductive Health for Youth Adults*, Focus on Young Adults, Pathfinder International: Washington, DC.

Shiraishi, S. (1997), *Young Heroes: Children, School, and the Politics of Family in New Order Indonesia*, Cornell University Southeast Asia Program: Ithaca.

Sikes, O.J. (1996), "Approaches to adolescent reproductive health: audience-specific strategies." *International Journal of Health Promotion and Education,* 3(3):15–17.

Singh, S. and Samara, R. (1996), "Early marriage among women in developing countries." *International Family Planning Perspectives*, 22(4):148–157, 175.

Singh, S. and Wulf, D. (1990), Today's *Adolescents, Tomorrow's Parents: A Portrait of the Americas*, Alan Guttmacher Institute: New York.

Slater, M.D. and Flora, J.A. (1991), "Health lifestyles: audience segmentation analysis for public health interventions." *Health Education Quarterly*, 18(2):221–233.

Smith, D.P. (1980), *Age at First Marriage. Comparative Studies. Cross National Summaries*, No. 7. World Fertility Survey: London.

Smith, P.C. (1980), "Asian marriage patterns in transition." *Journal of Family History*, 5(1):58–97.

Trussell, J., Menken, J. and Coale, A.J. (1979), "A General Model for Analyzing the Effects of Nuptiality on Fertility." In: *Nuptiality and Fertility*, L. Ruzicka, ed. Ordina Editions: Liege.

UNICEF. (1967), *Children and Youth in National Planning and Development in Asia. Volume 1. Policy and Planning*, UNICEF: Bangkok.

United Nations. (1983), *Marital Status and Fertility: A Comparative Analysis of World Fertility Survey Data for Twenty-One Countries*, United Nations: New York.

United Nations. (1988), "First marriage: patterns and determinants." Department of International Economic and Social Affairs. ST/ESA/SER.R/76. United Nations: New York.

United Nations. (1990), *Patterns of First Marriage: Timing and Prevalence*, United Nations: New York.

United Nations. (1994), *Programme of Action of the 1994 International Conference on Population and Development*, United Nations: New York.

United Nations. (1995), *World Population Prospects: The 1994 Revision*, United Nations: New York.

United Nations, ESCAP. (1989), *Review of Youth Policies in the ESCAP Region*, United Nations: New York.

Visaria, L. (1986), "Demography of the Indian youth." *Working Paper*, No. 4. The Gujarat Institute of Area Planning Working Paper Series.

Wallimann, I. and Zito, G.V. (1984), "Cohort size and youthful protest." *Youth and Society*, 16(1):67–81.

Watkins, S. (1986), "Regional patterns of nuptiality in Western Europe, 1870–1960." Chapter 8 In: *The Decline of Fertility in Europe: the Revised Proceedings of a Conference on the Princeton European Fertility Project*, A. Coale and S. Watkins, eds. Chapter 8. Princeton University Press: Princeton, NJ.

Weiner, M. (1971), "Political demography: an inquiry into the political consequences of population change." In: *Rapid Population Growth: Consequences and Policy Implications*, Edited by National Academy of Sciences. Chapter 15. Johns Hopkins Press: Baltimore.

Westoff, C.F., Blanc, A.K. and Nyblade, L. (1994), "Marriage and entry into parenthood." *Demographic and Health Surveys Comparative Studies*, No. 10. Macro International Inc.: Calverton, MD.

World Health Organization. (1993), *The Health of Young People: a Challenge and a Promise*, WHO: Geneva.

Wriggins, H.W. (1988), "Youth cohorts, population change and politics: five intervening variables." Paper prepared for a *Conference on Future Changes in Population Age Structure, Sopron, Hungary, Oct. 18–21.*

Wrigley, E.A. and Schofield, R.S. (1981), *The Population History of England 1541–1871*, Harvard University Press: Cambridge.

Xenos, P. (1997), "Measuring the Sexual System and Union Formation Among Adolescents in Developing Countries." Paper prepared for the *Workshop on Adolescent Sexuality and Reproductive Health in Developing Countries, Washington, DC, March 24–25.*

Xenos, P. (1990a), "The Social Demography of Asian Youth Populations: An Analysis of Projections to 2010." *Analysis of Population Trends and Projections in Asia 1980–2010*, East-West Population Institute: Honolulu. (Background Paper No. 3.)

Xenos, P. (1990b), "Youth, Sexuality and Public Policy in Asia: A Research Perspective." In: *The Family as an Asset: An International Perspective on Marriage, Parenthood and Social Policy*, S.R. Quah, ed. Chapter 14. Times Academic Press: Singapore.

Xenos, P., Achmad, S., Lin, H., Luis, P-K., Podhisita, C., Raymundo, C. and Thapa, S. (2000), "Asian youth in global context: indicators from Asian young adult reproductive health surveys." Paper prepared for the *International Symposium on Reproductive Health Research and Policy Issues of Adolescent and Unmarried Young Adults, Oct. 19–21, Shanghai*.

Xenos, P. and Gultiano, S.A. (1992), *Trends in Female and Male Age at marriage and Celibacy in Asia*, East-West Center Program on Population: Honolulu. (Sandra E. Ward. Papers of the Program on Population, No. 20.)

Xenos, P. and Gultiano, S.A. (2002), "The Marriage Ratios of Asian Countries and Cohorts." Paper presented at the *2002 IUSSP Regional Population Conference, June 10–13, Bangkok.*

Xenos, P. and Kabamalan, M. (1998a), "The changing demographic and social profile of youth in Asia." *Asia-Pacific Population Research Reports*, 12(Oct.):1–24.

Xenos, P. and Kabamalan, M. (1998b), "The social demography of Asian youth: A reconstruction over 1950–1990 and projection to 2025." The report on the project *Long-Term Transformations of Youth in Asia (Population Council Contract CI97.63A)*, Program on Population, East-West Center: Honolulu.

Part II

Policy Analysis, Models and Methods

CHAPTER 5. A FORMAL MODEL OF AGE-STRUCTURAL TRANSITIONS

NAN LI
Duke University

SHRIPAD TULJAPURKAR
Stanford University

Introduction

Population growth has long been a concern for high fertility countries. Few conclusions have been reached on such issues as interactions between population and economic development (Ehrlich 1968; Simon 1981), natural resources (Meadows et al. 1972; Cole et al. 1973) and environment changes (Cohen 1995). However, by 1989, 123 countries, representing 91 percent of the world's population, claimed to support family planning programs (United Nations 1989). Reducing fertility to the replacement level, about two children per woman, eventually would lead to stationary population, with constant size and age structure. Lowering fertility faster would result in a smaller size stationary population, producing greater space for high fertility countries to arrange relations between population and development. Reducing fertility faster, however, would lead to remarkable oscillations in the number of annual births. Further, the oscillations in births over time are not caused by oscillations in transition or in initial age structure. Even for fertility transitions as smooth as linear and initial populations whose age structures are as smooth as stable, remarkable oscillations in births can be observed. Over time, these oscillations in birth will manifest as peaks and valleys in the population age structure, leading to imbalances in the supply and demand of socioeconomic requirements such as education, employment and retirement.

How the size of the stationary population is determined by fertility transition, in which fertility reduces to replacement level, can be described by population momentum (Keyfitz 1971; Li and Tuljapurkar 1999). That the age structure of a stationary population is determined by mortality is also well known. However, in the process of fertility transition, how the age structure changes from its current shape to the stationary pattern, which we called the age-structure transition, has not yet been well addressed. Nevertheless, potential problems caused by the age-structure transition, which would occur shortly after the fertility transition begins, seem more important than those related to the stationary population, not only for governments, but the public as well.

Oscillations in births play an important role in age-structure transition. For what reasons and under what conditions does fertility transition cause such oscillations? Simulation methods,

S. Tuljapurkar et al. (eds.), Population, Resources and Development, 91–105.

such as cohort-component projections, are useful in demonstrating oscillation phenomena numerically with year-to-year calculations. However, in using these methods, the causes of population change are lost in the numerical calculations, and become invisible. In this paper, the analytical solution of the time-dependent population model (Li and Tuljapurkar, 2000) is used to analyze oscillations in age-structure transition.

The analytical solution shows that the general form of the change in the number of annual births, under demographic transitions, is the sum of a constant, plus declining sinusoidal waves. The frequencies of these sinusoidal waves are determined only by the constant mortality and fertility after the demographic transition, and are independent of the demographic transition and the initial population's age structure. In other words, oscillation is an intrinsic feature of age-structure transition, as these sinusoidal waves are determined only by how the post-transition mortality and fertility change over age, regardless of what the demographic transition and initial population may be. The question remains, under what conditions does the oscillation occur?

Both the initial age-structure and demographic transition can be arbitrary, and may produce changes in births ranging from monotonic curves to remarkable oscillations. On the other hand, the number of sinusoidal waves in the general solution is infinite; these waves may encompass any changes produced by the initial age-structure and demographic transition. To discuss under what conditions oscillation occurs, we first focus on the simplest form of the general solution. This form, which we call the simplest solution, is a constant plus a declining sinusoidal wave with the lowest frequency. We then discuss under what conditions this simplest solution best describes the changes in the number of annual births, which resolves when the oscillation occurs. We conclude that for stable initial population and linear fertility transitions, remarkable oscillation in births occurs when the period of transition is about half of the mean age of childbearing, and the higher the initial fertility level, the stronger the oscillation. Reducing the period of transition from half of the mean age of childbearing would enhance the oscillation, and vice versa.

The above conclusions derive from analyses of the general and simplest solutions, without using population data. At the end of this paper, we use data from the less and least developed regions, provided by the United Nations (1998), to check and illustrate these conclusions.

The Solution of the Time-dependent Population Model

Denote by α and β the minimum and maximum fertile ages, $b(t)$ the number of female births at time t, $\phi\ (x,t)$ the maternity function of age x at time t, and $g(t)$ the number of female births at time t produced by initial population. The time-dependent population model can be written as

$$b(t) = \int_{\alpha}^{t} \phi(x,t)b(t-x)dx + g(t). \tag{1}$$

When the maternity function $\phi\ (x,t)$ changes in a finite period γ, it can be written as

$$\phi(x,t) = \phi_F(x), t \geq \gamma, \tag{2}$$

where $\phi_F(x)$, constant over time t, denotes the maternity function after the demographic transition that lasts γ years, and the subscript 'F' stands for 'final'. When the transition time, γ, is less than β years, the general solution of (1) is found (Li and Tuljapurkar, 2000) as

$$b(t) = \sum_{n=0}^{\infty} a_n \exp(r_n t), \tag{3}$$

where r_n is the solution of equation

$$\int_{\alpha}^{\beta} \phi_F(x) \exp(-r_n x) dx = 1. \tag{4}$$

Let

$$\mu_n = \int_{\alpha}^{\beta} x \phi_F(x) \exp(-r_n x) dx, \tag{5}$$

where μ_0 is the average age of childbearing in the final stationary population. Coefficients a_n are given as

$$a_n = \int_{0}^{\beta} G(x) \exp(-r_n x) dx / \mu_n, \tag{6}$$

where

$$G(t) = g(t) + \int_{\alpha}^{t} [\phi(x,t) - \phi_F(x)] \sum_{n=0}^{N} g_n(t-x) dx, N = \text{int}(\gamma/\alpha), \tag{7}$$

$$g_n(t) = \int_{\alpha}^{t} \phi(x,t) g_{n-1}(t-x) dx, n \geq 1, \tag{8}$$

$$g_0(t) = g(t). \tag{9}$$

When $G(x)$ in (6) is substituted by $g(x)$, solution (3) reduces to the form that is used in virtually all demographic textbooks to describe the situation only of constant fertility and mortality. In fact, however, solution (3) describes the change in births under any transition of vital rates that is completed in less than β years, if using $G(x)$ in (6).

Solution (3) shows that oscillation is an intrinsic feature of population dynamics under demographic transition, because the frequencies of the sinusoidal waves, which are the imaginary part of r_n, are determined in (4) only by $\phi_F(x)$, regardless of whatever the ϕ (x,t) and $g(t)$ may be.

Simple Situations

To discuss how the amplitudes of these sinusoidal waves, a_n, are determined, we must simplify formula (6), which is too complicated for analysis. To do so, we choose a simple but representative situation, which includes stable initial population and linear fertility transition. In situations in which the initial population is stable,

$$p(x,0) = l_x \exp(-rx), \tag{10}$$

where l_x is the survival rate from birth to age x, and r, the growth rate of the stable population, satisfies

$$\int_{\alpha}^{\beta} \phi(x,0) \exp(-rx) dx = 1. \tag{11}$$

When only fertility level changes linearly to replacement in less than α years,

$$\phi(x,t) = \left[R_0 - \frac{R_0 - 1}{\gamma} t \right] \phi_F(x), t \leq \gamma \leq \alpha, \tag{12}$$

where R_0 is the initial net reproduction rate

$$R_0 = \int_{\alpha}^{\beta} \phi(x,0) dx. \tag{13}$$

In the above situations, coefficients a_n are simple:

$$a_n = \frac{R_0 - 1}{R_0 \mu_n (r - r_n) \gamma} \int_0^{\gamma} \exp[(r - r_n)x] dx. \tag{14}$$

This expression of a_n is exact for $\gamma \leq \alpha$ and accurate for $\alpha < \gamma \leq \beta$ (Li and Tuljapurkar, 1999).

The Simplest Situation

To discuss under what conditions oscillation occurs, we must reduce the number of sinusoidal waves in the solution, since the combination of infinite waves is beyond understanding. To do so, we discuss the simplest solution, which contains only one sinusoidal wave, with the lowest frequency.

Denote the simplest solution as

$$b_s(t) = a_0 + a_1 \exp(r_1 t) + \overline{a}_1 \exp(\overline{r}_1 t), \tag{15}$$

where the subscript 's' stands for 'simplest'. We look for values of γ that make $b_s(t)$ close to $b(t)$. Since

$$0 = r_0 > \mathrm{Re}(r_1) > \ldots > \mathrm{Re}(r_{n-1}) > \mathrm{Re}(r_n), \tag{16}$$

where $\mathrm{Re}(r_n)$ is the real part of r_n, $b_s(t)$ is naturally accurate for large t. Therefore the task is to seek conditions which make $b_s(t)$ accurate for small t. This is obviously to find γ to

$$\min[b(0) - a_0 - 2\mathrm{Re}(a_1)]^2 = \min[1 - a_0 - 2\mathrm{Re}(a_1)]^2. \tag{17}$$

Let the real and imaginary part of r_1 be u_1 and v_1, respectively:

$$r_1 = u_1 + iv_1, \tag{18}$$

where v_1 is the lowest frequency and u_1 is the decline rate of the sinusoidal wave with the lowest frequency. Assuming (Pollard 1973: 31) that $\phi_F(x)$ satisfies normal distribution $N(\mu_0,\sigma^2)$ and noticing that r and $|u_1|$ are small variables, where $r \approx \frac{\log(R_0)}{\mu_0}$ (Pollard 1973: 31), we have

$$a_0 + 2\mathrm{Re}(a_1) \approx \frac{R_0 - 1}{R_0}\left[\frac{1}{\log(R_0)} + \frac{2}{\mu_0 v_1}\frac{1 - \cos(v_1\gamma)}{v_1\gamma}\right]. \tag{19}$$

Therefore, the task is to find the value of γ, which makes

$$\frac{d}{d\gamma}\left[\frac{1 - \cos(v_1\gamma)}{v_1\gamma}\right] = 0. \tag{20}$$

Equation (20) yields $v_1\gamma \approx \pi$. For human populations, $v_1 \approx \frac{2\pi}{\mu_0}$ (Pollard 1973: 31), thus

$$\gamma \approx \frac{\pi}{v_1} \approx \frac{\mu_0}{2}. \tag{21}$$

The values of γ in (21) make $b_s(t)$ close to $b(t)$. How close they can be depends on the value of μ_0. Since the typical values of μ_0 in human populations are

$$\mu_0 \approx 2\alpha, \tag{22}$$

the values of γ in (21) will make $b_s(t)$ fit $b(t)$ well. Why? Fertility decline causes $b(t)$ to decline from $t = 0$ until the transition stops at $t = \gamma$. Then $b(t)$ increases as the larger cohorts born before $t = 0$ reach childbearing, if (22) holds. This increase of $b(t)$ will stop at $t \approx \mu_0$, as these larger cohorts reach age μ_0, after which $b(t)$ decreases. Thus the length of the first cycle in the oscillation of $b(t)$ is about μ_0, which is just the cycle of $b_s(t)$. In other words, if the period of transition is close to the half cycle of the sinusoidal wave with the lowest frequency, the first half cycle of this sinusoidal wave will be generated. Further, if the starting age of childbearing is also close to this half cycle, the second half cycle of this sinusoidal wave is also generated. Thus, this wave will maintain itself, because its first cycle is generated and its frequency is the one determined by $\phi_F(x)$.

Maximizing $\mathrm{Re}(a_1)$ also minimizes $\mathrm{Im}(a_1)$, the imaginary part of a_1, thus the simplest solution for $\gamma \approx \frac{\mu_0}{2}$ can be written as

$$b_s(t) = a_0 + 2\mathrm{Re}(a_1)\exp(u_1 t)\cos\left(\frac{2\pi}{\mu_0}t\right). \tag{23}$$

The value of a_1 given by (14) is for the situation that all terms in (3) are involved. Using the simplest solution, there are many choices of a_1 and the simplest one is to fit $b(0)$. Therefore the simplest solution can be rewritten as

$$b_s(t) = a_0 + (1 - a_0)\exp(u_1 t)\cos\left(\frac{2\pi}{\mu_0}t\right). \tag{24}$$

For $\gamma \neq \frac{\mu_0}{2}$, oscillation terms other than $\cos(\frac{2\pi}{\mu_0}t)$ should be used in (24) in order to obtain better descriptions of $b(t)$. For the values of γ being close to $\frac{\mu_0}{2}$, adding the term $\sin(\frac{2\pi}{\mu_0}t)$ is the simplest choice, which introduces only one phase, θ, into (24). This phase should make the first minimum value of $b_s(t)$ at $t = \gamma$. Therefore

$$\theta = \pi - \frac{2\pi}{\mu_0}\gamma, \tag{25}$$

and thus the simplest solution can be rewritten as

$$b_s(t) = a_0 + A_1\exp(u_1 t)\cos\left(\frac{2\pi}{\mu_0}t + \theta\right), \ A_1 = \frac{1 - a_0}{\cos(\theta)}, \tag{26}$$

and (24) is a special case of (26) for $\gamma = \frac{\mu_0}{2}$.

The simplest solution concludes that remarkable oscillation in birth occurs when the period of transition is about half of the mean age of childbearing, for situations of linear fertility transition and stable initial population.

Properties of the Simplest Solution

The typical $b(t)$ from a linear fertility transition in about α years, as described by $b_s(t)$, would be an oscillation with a cycle of μ_0 years. Because the typical situations of fertility transition are $R_0 > 1$, and

$$a_0 = \frac{R_0 - 1}{R_0\mu_0 r}\frac{\exp(r\gamma) - 1}{r\gamma} \approx \frac{2(R_0 - 1)(\sqrt{R_0} - 1)}{R_0[\log(R_0)]^2} \approx \frac{2}{R_0(\sqrt{R_0} + 1)} < 1, \tag{27}$$

thus A_1 would be positive and the typical $b(t)$ would reach maximum values at $t = 0, \mu_0, 2\mu_0, \ldots$. Since $u_1 < 0$, thus the oscillation declines over time and the first cycle is the most significant one, which produces a baby bust. The first cycle, in which $b(t)$ declines from 1 to the minimum at about γ years, would yield a baby bust with the maximum amplitude of

$$b(0) - b(\gamma) = (1 - a_0)\left(1 + \frac{\exp(u_1\gamma)}{\cos(\theta)}\right). \tag{28}$$

From (27), the larger the R_0, the smaller the a_0, hence the more remarkable the baby bust. On the other hand, because

$$a_0 = \frac{R_0 - 1}{R_0\mu_0 r}\left(1 + \frac{1}{2}r\gamma\right), \frac{\partial a_0}{\partial \gamma} > 0, \tag{29}$$

and when $\gamma \approx \frac{\mu_0}{2}$,

$$\frac{\partial}{\partial \gamma}[b(0) - b(\gamma)] \approx -[1 + \exp(u_1\gamma)]\frac{\partial a_0}{\partial \gamma} + u_1 \exp(u_1\gamma)(1 - a_0) < 0. \tag{30}$$

Thus reducing γ would also yield a more remarkable baby bust.

When the initial population is stable and the period of linear fertility transition is close to half of the mean age of childbearing, the above properties show that there would be a remarkable oscillation and that its first cycle would be a baby bust; the higher the initial fertility level (R_0) or the shorter the transition period (γ), the stronger the oscillation or the more significant the baby bust.

Examples and Discussion

The above conclusions are based on the simplest solution, without involving any specific data. We now use the data of the less and least developed regions, provided by the United Nations (1998), to check and illustrate these conclusions, because the fertility levels of the two regions are higher than replacement level.

The United Nations divides the world into more, less and least developed regions (United Nations 1998). Populations in these regions were 1.171, 3.938 and 0.579 billions respectively in 1995. For each region, the United Nations provided age-specific birth rates for 1990–1995 and population for 1995. For mortality, there are only values of life expectancy at birth, e_0, for 1990–1995. We use data from the less and least developed regions to explore the results of fertility transition. For mortality data, we first choose age-specific death rates from the Coale-Demeny (Coale and Demeny 1983) life tables for Western female models with $e_0 = 65$, and 50 for the less and least developed regions, respectively. We then adjust these death rates using proportional changes to fit the reported values of e_0, for each of the two regions. Parameters for the two regions are shown in Table 1.

The age structures of both stable and actual population for the less and least developed regions are shown in Figures 1a and 1b, respectively. The stable age structure of the least developed region is much younger than that of the less developed one

Table 1. Parameters of less and least developed regions.

	e_0 (Female life expectancy at birth)	R_0 (Net reproduction rate)	μ_0 (Average age of childbearing)	r (Growth rate of stable population)	u_1 (Decline rate of the lowest-frequency wave)	v_1 (Lowest frequency)
Less developed region	63.7	1.43	27.4	0.013	−0.0446	0.232
Least developed region	50.8	1.98	28.4	0.024	−0.0449	0.210

because its fertility level, measured by the value of R_0, is much higher, as can be seen in Table 1.

Even for fertility transitions as smooth as linear and initial populations whose age structures are as smooth as stable in both regions, remarkable oscillations in births are observed in Figures 2a and 2b when the periods of transition are about half of the mean ages of

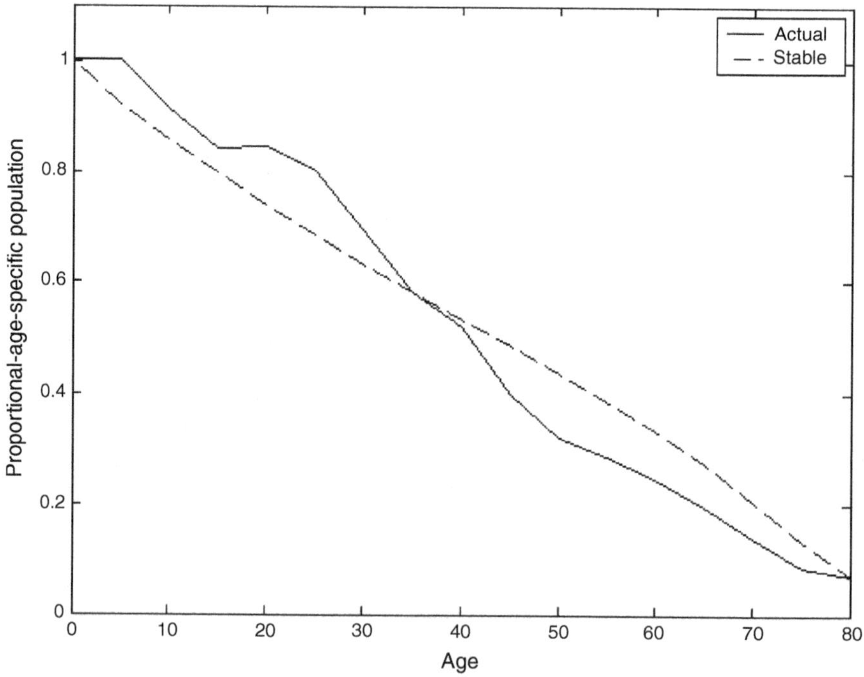

Figure 1a. Initial age structures of less developed countries in 1995.

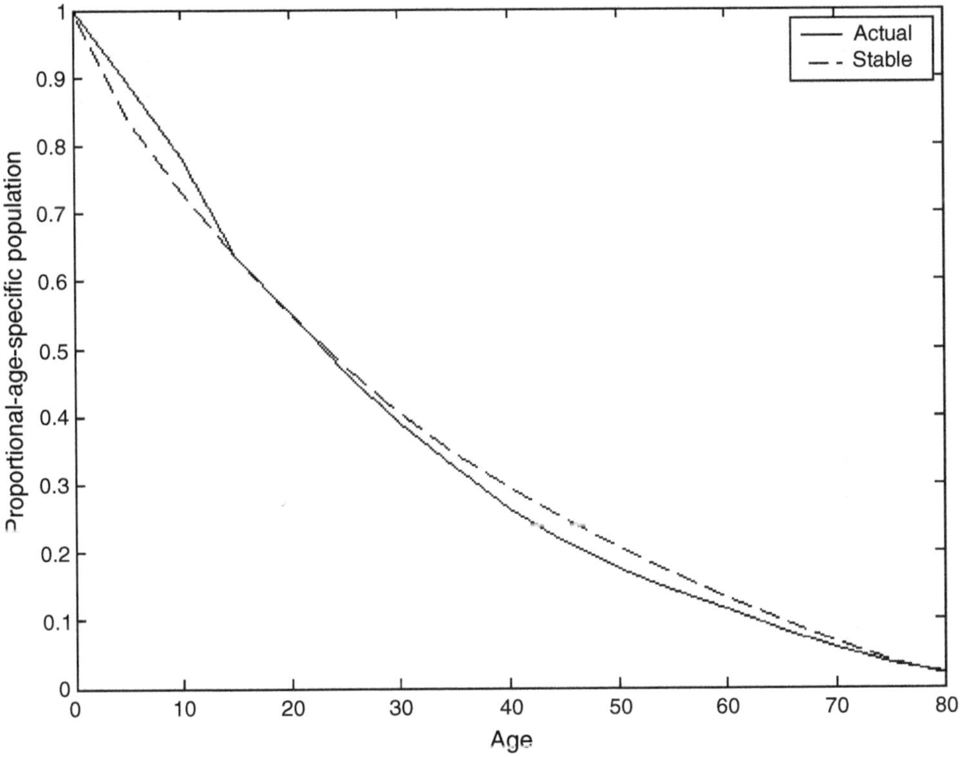

Figure 1b. Initial age structures of least developed countries in 1995.

childbearing, $\gamma = 15$ years. The cycles of these oscillations are about 30 years, the mean age of childbearing in the two regions as predicted by (21). In both regions, the first cycle produces a baby bust as specified by (27) and (28). Figures 2a and 2b also confirm that the higher the initial fertility the stronger the oscillation, as can be seen by the oscillation of the least developed region, which is stronger than the less developed one, and its value of R_0 is higher.

Since there is a larger population at younger ages in the actual age structures than the stable ones, as can be seen in Figures 1a and 1b, the births of the former would be more than the latter for both regions. This explains why the births from simulation, using actual age structure, are greater than those from using stable ones, as shown in Figures 2a and 2b. Although the simplest solution does not describe the births from simulation using actual age structure as well as that of using stable age structure, it still captures the amplitudes and cycles well. In general, the closer the actual age structure to the stable one, the better the simplest solution describes the dynamics of births.

Figures 3a and 3b show the results of changing γ from 15 years to 10 and 20 years, for the less and least regions respectively. They confirm that faster transition ($\gamma = 10$) produces stronger oscillations as predicted by (30), and vice versa ($\gamma = 20$).

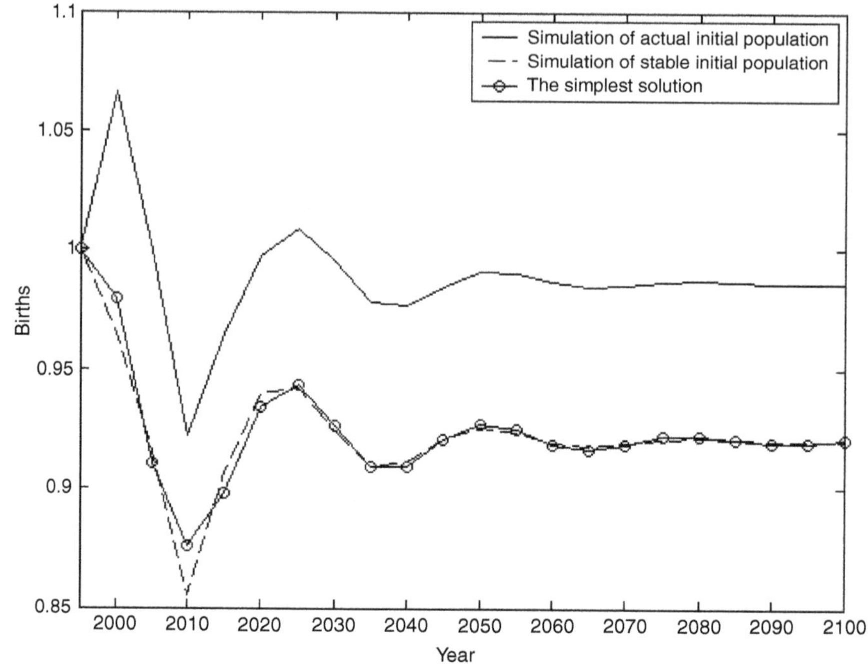

Figure 2a. Birth sequences of less developed countries, $\gamma = 15$ years.

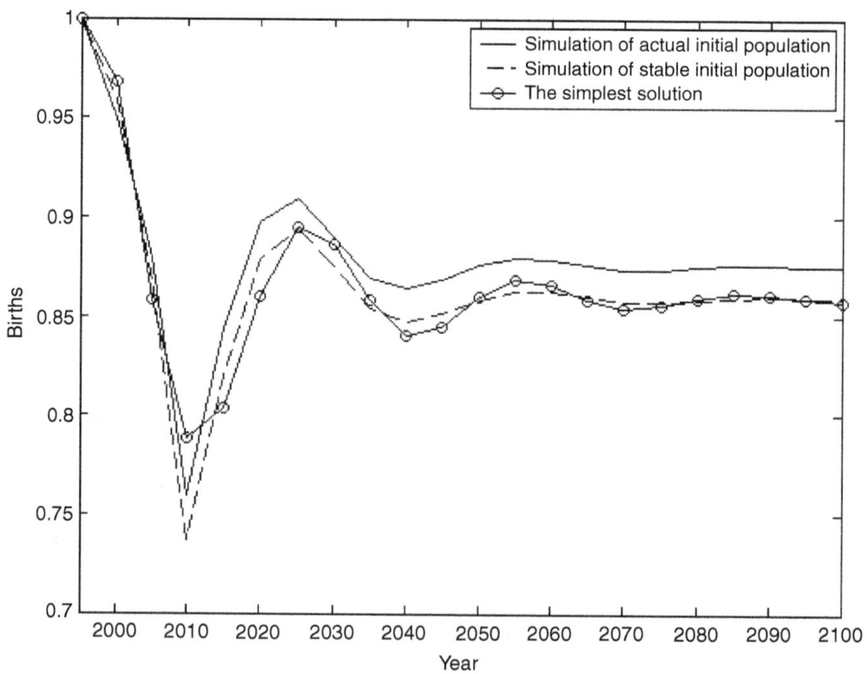

Figure 2b. Birth sequences of least developed countries, $\gamma = 15$ years.

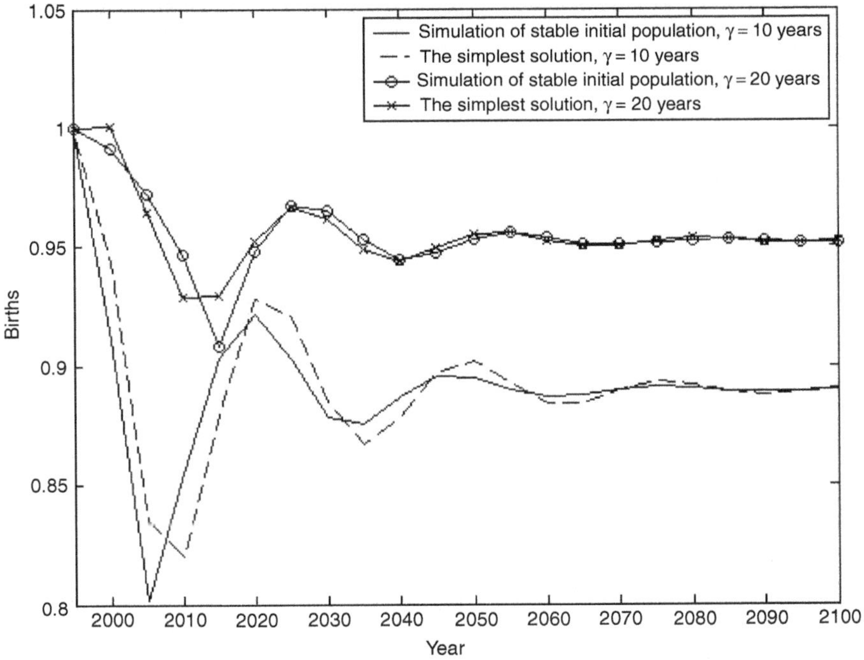

Figure 3a. Birth sequences of less developed countries.

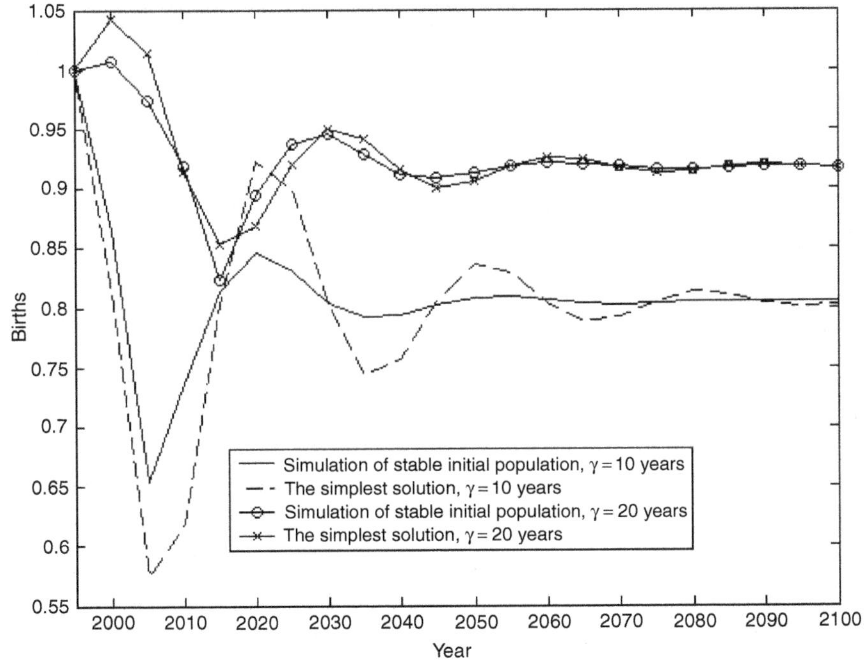

Figure 3b. Birth sequences of least developed countries.

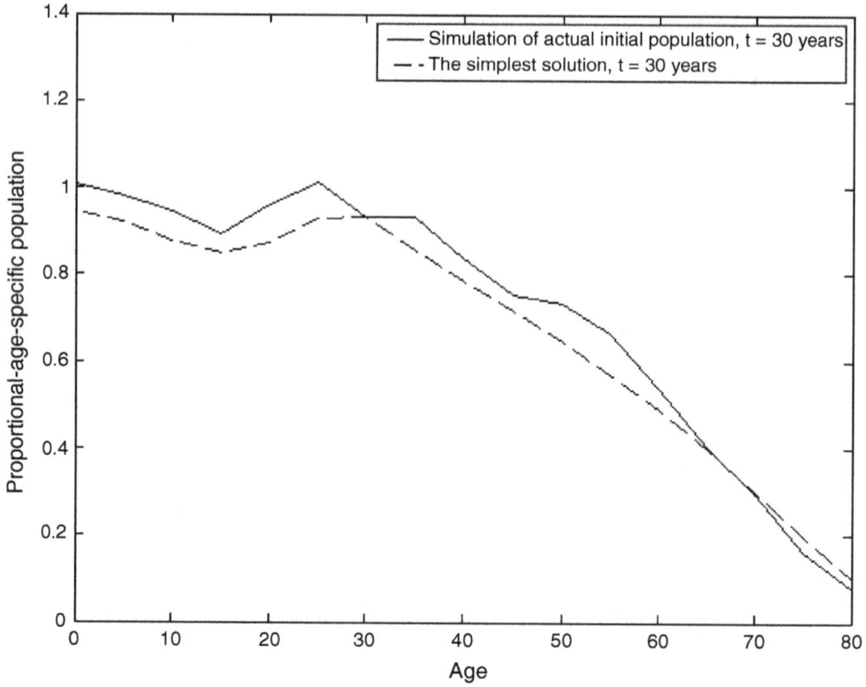

Figure 4a. Age structures of less developed countries, $\gamma = 15$ years.

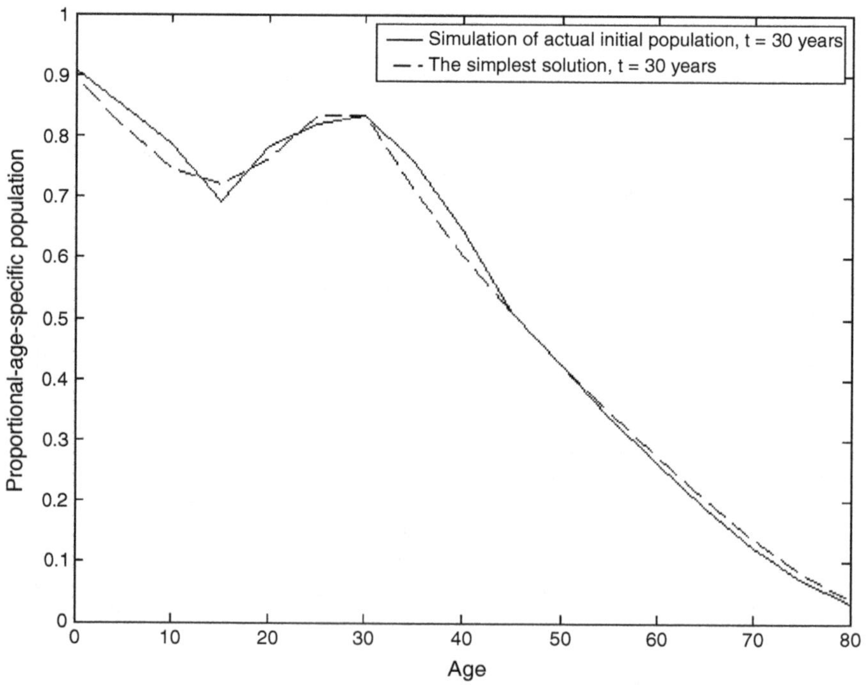

Figure 4b. Age structures of least developed countries, $\gamma = 15$ years.

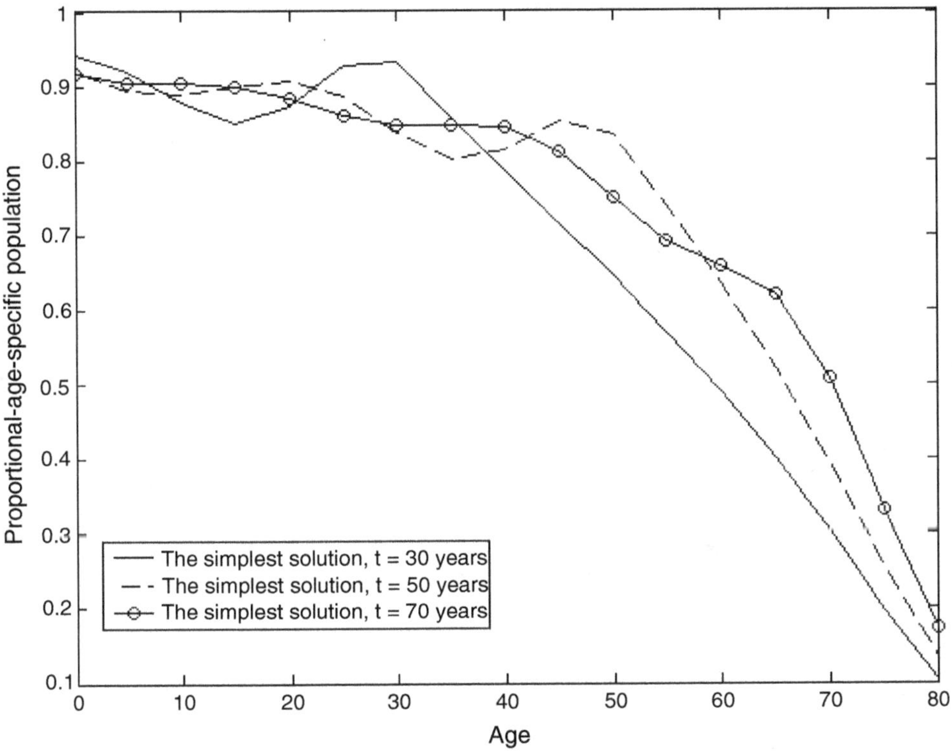

Figure 5a. Age structures of less developed countries, $\gamma = 15$ years.

Figures 4a and 4b show the peaks and valleys in the age structure 30 years after the beginning of the fertility transition, for the less and least regions respectively. The peaks and valleys across age result from the oscillation in births over time. The difference between the actual and stable age structures, however, does not remarkably vary the oscillation patterns in birth sequences and hence in age structures, as shown in Figures 4a and 4b. Again, this is due to the actual age structure being close to the stable one.

Figures 5a and 5b indicate the age structure transition by showing the age structures at 30, 50 and 70 years after the beginning of the fertility transition, for the less and least regions respectively. In the first 30 years, a baby bust would reduce young dependants, as shown in the solid curves. After 50 years, however, the baby bust would reduce the working age population, as drawn in dashed lines. After 70 years, the baby bust would be levelled off by high mortality in older ages, as can be seen from the dotted curves.

Conclusion

These examples confirm our conclusions regarding why and under what conditions oscillations in births occur. Oscillation is the intrinsic feature of the age structure transition, because the values of v_1 in Table 1, the lowest frequency of the sinusoidal waves in the

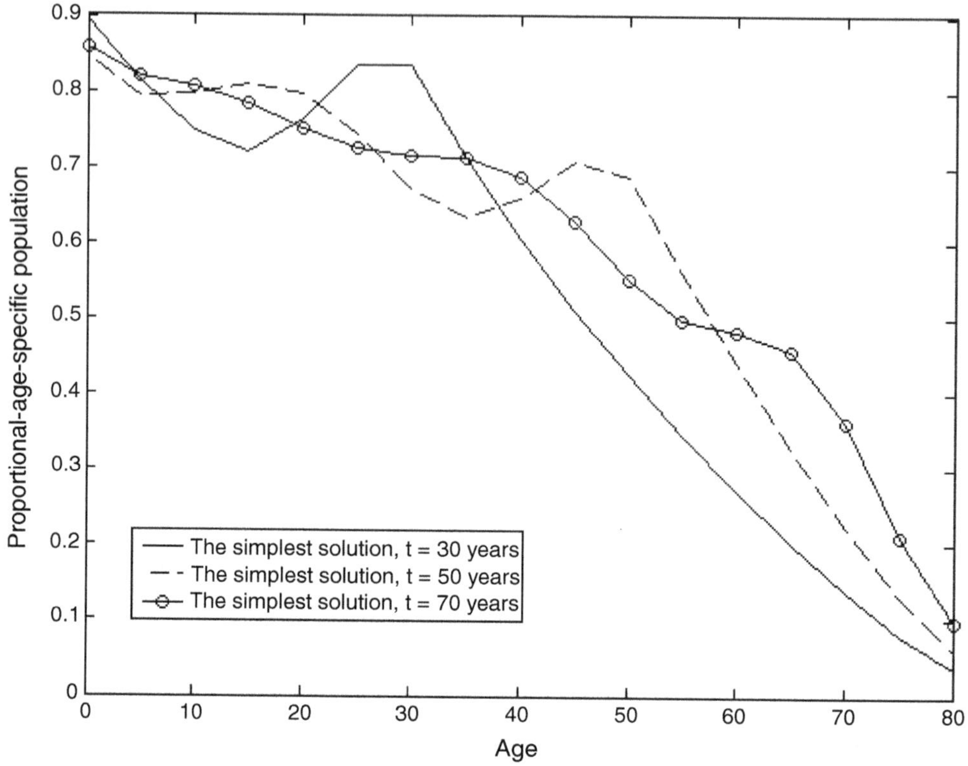

Figure 5b. Age structures of least developed countries, $\gamma = 15$ years.

general solution, are determined only by the post-transition mortality and fertility in (4), regardless of whatever the demographic transition and initial population may be. Remarkable oscillation would occur when the period of transition is about half of the mean age of childbearing, even for initial populations as smooth as stable and fertility transitions as smooth as linear, as confirmed by Figures 2a and 2b. All other things being equal, oscillation would be stronger for higher fertility countries, as noted by comparing Figures 2a and 2b, and also for faster fertility transitions, as indicated in Figures 3a and 3b. Oscillations in birth over time will manifest into a baby bust in the age structure of population, as shown in Figures 4a and 4b. These baby busts may lead to imbalances in the supply and demand of such socioeconomic needs as education, employment and retirement, as implied by Figures 5a and 5b.

References

Coale, A. and Demeny, P. (1983), *Regional Model Life Tables and Stable Population*, Academic Press: New York.

Cohen, J.E. (1995), *How Many People Can the Earth Support?* W.W. Norton: New York.

Cole, H.S.D., Freeman, C., Jahoda, M. and Pavitt, K.L.R. (1973), *Models of Doom: A Critique of The Limit to Growth, With a Replay by the Authors of The Limit to Growth*, Universe Books: New York.

Ehrlich, P. (1968), *The Population Bomb*, Ballantin Books: New York.

Keyfitz, N. (1971), "On the momentum of population growth." *Demography*, 8:71–80.

Li, N. and Tuljapurkar, S. (1999), "Population momentum for gradual demographic transitions." *Population Studies*, 53:255–262.

Li, N. and Tuljapurkar, S. (2000), "The solution of time-dependent population models." *Mathematical Population Studies*, 7(4):311–329.

Meadows, D.H., Meadows, D.L., Randers, J. and Behrens, W.W. (1972), *The Limit to Growth: A Report for the Club of Rome's Project on the Predicament of Mankind*, Signet, New American Library: New York. (2nd Edition, 1974.)

Pollard, J. (1973), *Mathematical Models of the Growth of Human Population*, Cambridge University Press: Cambridge.

United Nations. (1989), *Trends in Population Policy*, United Nations: New York.

United Nations. (1998), *World Population Prospects: The 1996 Revision*, Population Division, United Nations.

CHAPTER 6. PROJECTING NUMBERS OF LIVING CHILDREN OF OLD PEOPLE, WITH EXAMPLES FROM KOREA AND THAILAND

JOHN BRYANT
Institute for Population and Social Research, Mahidol University

Introduction

Demographic change at the level of the population is inevitably accompanied by demographic change at the level of the family. Population ageing, for instance, entails a reduction in the average number of living children of old people. The family-level changes may turn out to be just as important as the population-level ones. It is possible, for instance, that present-day systems of family-based old age support will be weakened when responsibilities are shared among a much smaller number of children. But whereas articles on ageing or old people are usually replete with statistics on projected numbers of old people, projected ratios of workers to retirees, and other population-level trends, few try to quantify future family-level trends. Few articles, for instance, try to put a figure on the number of childless elderly 10 or 20 years into the future.

Methods for projecting family-level characteristics exist. Micro-simulations of kinship patterns such as SOCSIM and CAMSIM can be used to give detailed analyses of possible futures, as Zeng (1986) and Jiang (1994) have done for China. A macro-simulation developed by Zeng et al. (1999) can produce much of the same detail as the micro-simulations, using more readily-available data. Such data are, nevertheless, not available in many developing countries.

This paper describes a simple macro-simulation model developed by the author for projecting the distribution of elderly women by numbers of living children 25 years into the future. The model is designed for late-transitional and post-transitional populations. For most developing countries, all the input data can be found in published tabulations. The model is implemented on a computer spreadsheet, a copy of which may be obtained from the author. This paper describes the model and a test of the model's validity. It then applies the model to South Korea's demographic transition and to Thailand's HIV/AIDS epidemic.

The Model and its Implementation

Consider a woman in her late thirties belonging to a late-transitional or post-transitional population who is trying to predict the number of living children she will have in her

S. Tuljapurkar et al. (eds.), Population, Resources and Development, 107–117.

sixties. Since fertility and mortality rates are low, the chance that the woman will have another birth and the chance that one of her children will die before she reaches 60 are both small. The woman can therefore treat the number of children she has now as a good approximation of the number she will have then. For a woman currently in her fifties, the approximation is even better.

The same is essentially true for a demographer trying to predict numbers of living children for a whole cohort. If the cohort belongs to a late-transitional or post-transitional population, and has almost passed out of the child-bearing ages, then the demographer can treat the current distribution of cohort members by numbers of living children as a good approximation of the distribution in two or three decades' time. If the demographer also makes sensible adjustments for fertility and mortality then the approximation should be quite close – or at least as close as it can be, given uncertainties about future mortality and fertility levels, and given inaccuracies in the data.

The limitations imposed by inaccuracies in the data should not be underestimated. Table 1 shows estimates of mean numbers of living children in Thailand taken from censuses and surveys. Some of the inconsistencies apparent in the data can be attributed to differences in definitions (see the table note), but some must also be attributed to data error. Though

Table 1. Mean number of living children to Thai women, various sources.

Source	Year	Age of women		
		60–69	70+	60+
Census	1980	5.2	5.0	–
Survey of population change	1984	5.1	4.8	–
SECAPT* survey	1986	–	–	5.1
Survey of population change	1989	4.4	3.7	–
Census	1990	4.0	3.8	–
Survey of population change	1995	4.7	4.5	–
Survey of the welfare of elderly in Thailand	1995	–	–	5.3

*Social and Economic Consequences of the Ageing of the Population of Thailand.

NOTES. As Chayovan, Knodel, and Siriboon (1990: 47–9) point out, results for numbers of living children in the original 1980 census report are questionable because tabulators, probably inadvertently, treated women with unknown numbers of children as having no children. I have cited the results which Chayovan et al. calculated from the one percent sample. The tabulators avoided making the same mistake in 1990.

The SECAPT survey and the Survey of the Welfare of Elderly in Thailand (SWET) both explicitly referred to own children, adopted children, and stepchildren, whereas the censuses and surveys of population change left the exact meaning of 'children' undefined. This is probably one reason why the SECAPT survey and the SWET obtained higher estimates (Knodel and Chayovan 1997: 54).

SOURCES. Censuses: Chayovan, Knodel, and Siriboon (1990: Table 16); National Statistical Office (1992: Table 13); SPC: NSO (1986: Table 6; 1989: Table 8; 1997; Table 6); SECAPT: Knodel, Chayovan, and Siriboon (1992: Table 2); SWET: Chayovan and Knodel (1996: Table 2.15).

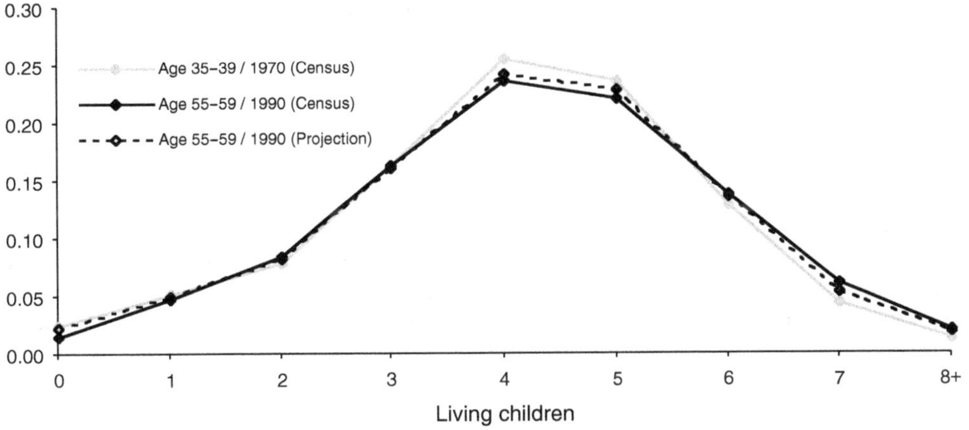

Figure 1. Distribution of South Korean women by numbers of living children.
Source: Census results calculated from UN Population Division (2000: Table 9). Projection results generated by model.

seldom mentioned in the literature on ageing, the existence of errors in old people's reports about children is a standard assumption in indirect methods for estimating fertility.

However, the data for South Korea are much more consistent, and presumably more reliable, than those for Thailand. Figure 1 uses these data to illustrate changes over time in a cohort's distribution by numbers of living children. The figure shows results from the 1970 census for ages 35–39, and from the 1990 census when the same cohort is aged 55–59. The figure also shows the projection model's prediction for 1990, based on the 1970 data. The reported distribution at ages 35–39 is clearly a good approximation to the reported distribution at ages 55–59. Allowing for a small net increase in numbers of children during the intervening years, which the projection does, gives an even better approximation.

The most substantively important discrepancy between the projected results and the 1990 census results concerns women with no living children. The model predicts that 2.3 percent of women should have no living children, while the census reports only 1.6 percent having no living children. Despite the high quality of the Korean data, the discrepancy may owe something to data error. The subtotals in the 1990 census table for numbers of living children reported by the UN Population Division (2000: Table 8) do not quite add up to the reported total, and this total is slightly different from that given in other census tables. The Population Division's notes to the table include the comment "Reason for discrepancy between these figures and corresponding figures shown elsewhere not ascertained".

By repeating the same calculations for many cohorts, and then aggregating the results, it is possible to trace how the distribution by number of living children changes for the various age groups of interest. This strategy of projecting cohorts and then aggregating is essentially the same as that used by classical population projection methods to trace out future changes in age–sex structure. The most important difference between the two projection methods is that the one described here is multi-state, whereas classical projections are single-state. Classical projections are able to represent the characteristic of interest – the size of a particular age–sex category – with a single number. The method described here must use several numbers: one for people with no living children, one for people with one child, and so on.

In a single-state projection the calculations required to advance a cohort forward one projection interval are straightforward. If a cohort has 1000 members at the beginning of an interval, and the probability that a randomly chosen member will survive until the end of the interval is 0.99, then the size of the cohort at the end of the interval is calculated as follows:

$$(0.99)(1000) = 990 \tag{1}$$

The projection method used in this paper, being a multi-state, must use a transition matrix rather than a single probability. Equation 2 is a simplified example of the calculations required. The vector v holds the numbers representing the state of the cohort at the beginning of the period. In Equation 2, 10 percent of the cohort have no children, 20 percent have one child, 40 percent have two children, and 30 percent have three children. The transition matrix M describes how demographic processes operating during the projection interval affect the distribution of cohort members by numbers of living children. The vector w is the outcome of multiplying v by M, and shows the cohort's distribution at the end of the interval.

$$\overset{\mathbf{M}}{\begin{bmatrix} 1 & 0.1 & 0.01 & 0.001 \\ 0 & 0.9 & 0.18 & 0.027 \\ 0 & 0 & 0.81 & 0.243 \\ 0 & 0 & 0 & 0.729 \end{bmatrix}} \overset{\mathbf{v}}{\begin{bmatrix} 0.1 \\ 0.2 \\ 0.4 \\ 0.3 \end{bmatrix}} = \overset{\mathbf{w}}{\begin{bmatrix} 0.1243 \\ 0.2601 \\ 0.3969 \\ 0.2187 \end{bmatrix}} \tag{2}$$

Each of the four columns of M indicates the end-of-interval distribution associated with a different start-of-interval number of children. The first column shows the end-of-interval distribution for people starting the interval with no children, the second column shows the distribution for people starting with one child, and so on. Because of the particular way in which matrix multiplication is defined, multiplying v by M is the same as taking a weighted sum of the columns of M, with the weight for the first column being the first number in v, the weight for the second column being the second number, and so on. In other words, matrix multiplication sums up the end-of-interval distributions for the four groups, weighting each distribution by the relative size of each group.

The projection method described here uses two types of transition matrix. One type deals with children who have already been born at the start of each projection interval. There is a different matrix of this type for every age group for every five-year interval. The matrix for any particular age group and interval is calculated from the average probability that, for a woman in the age group at the start of the interval, a randomly chosen child survives until the end of the interval. This probability is calculated from the distribution of living children by age and the survival probabilities at each age. The model treats deaths as independent, which permits it to use binomial distributions for the end-of-interval distributions. The matrix M in Equation 2 is an example of this type of matrix; M was calculated using a survival probability of 0.9.

The other type of transition matrix used by the projection model deals with children born during each interval. The model uses three such matrices for the cohort aged 35–39 at the start of the first projection interval: one for each interval in which some or all of the cohort's members are below the maximum age of reproduction. The model uses two matrices for the cohort aged 40–44 at the start of the first interval, and one for the cohort aged 45–49. The method used for calculating the matrices relies on an assumption that no woman gives birth to more than one child during a single projection interval of five years. This assumption comes fairly close to being met when the model is applied to late-transitional or post-transitional populations, since the cohorts being projected are near the end of their reproductive span, and fertility in late-transitional and post-transitional populations is low. Once this assumption has been made, the number of births during the interval, which is calculated from age-specific fertility rates, can be reinterpreted as the probability of giving birth. Multiplying this by the probability that a newborn baby will survive until the end of the interval gives the probability that a woman has one extra living child at the end of the interval. This value and its complement are placed in the matrix columns to represent the end-of-interval distributions for children born during the interval. The first column of the matrices, showing end-of-interval distributions for women with no children has been subjected to a further adjustment, based on the assumption that women who are never married will not be able to marry and have children before the end of their reproductive periods. The spreadsheet in which the model has been implemented lets users adjust fertility and mortality rates and introduce some correlation between deaths in the same family. Users can easily compare the results generated under the adjusted parameters with those generated under the original ones. The spreadsheet is heavily annotated and provides large numbers of graphs and tables for the projection results.

One of the main principles guiding the implementation of the model is that the input data should be easy to obtain. Users do not need to supply life tables for the projection period; instead they simply enter the median variant age-structures shown in the United Nations Population Division's World Population Prospects and the spreadsheet calculates survival probabilities from these. Users do need to supply some age-specific fertility rates, but these are usually readily available. The only data required by the model which can be difficult to find are data on the distribution of women by numbers of living children. Although reports from censuses and national surveys in developing countries generally supply the necessary tabulations, reports from developed countries generally do not. Ideally, the projection would have considered the children of both men and women. Data on male fertility are, however, so scarce that men have been left out of the current version.

Validating the Model

Census results for South Korea are the only data on numbers of living children I have been able to find which seem reasonably reliable and which extend over a sufficiently long period to allow the model to be tested. Figure 1 has already shown that the fit

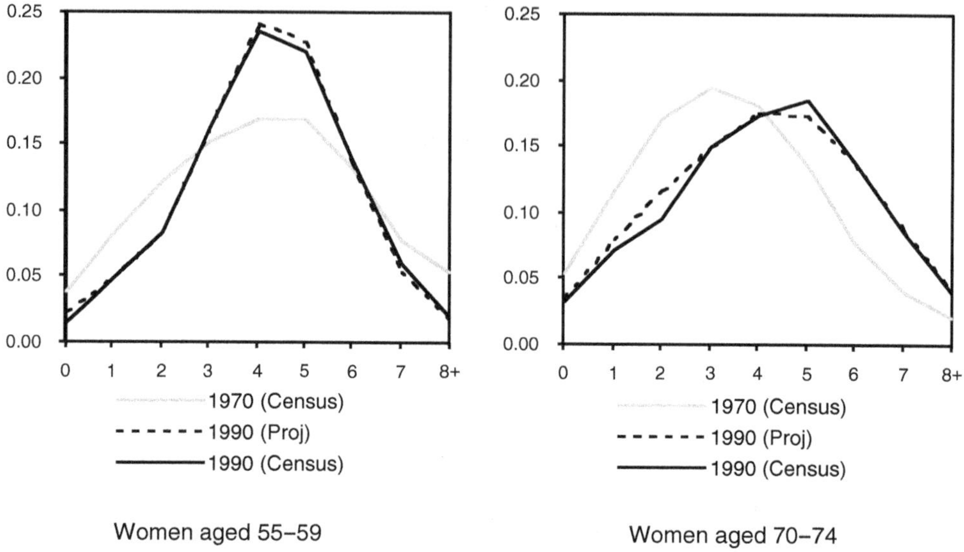

Figure 2. Distribution of South Korean women by numbers of living children.

between the projected and census figures is quite good. Figure 2 shows the distribution of South Korean women aged 55–59 and aged 70–74 by numbers of living children, as reported by the census in 1970 and in 1990. It also shows distributions in 1990 as predicted by the projection model from 1970 data and from demographic rates over the intervening period. The correspondence with the 1990 census results is not perfect. However, given the likelihood of small errors in both the 1970 and 1990 census data, the fit seems good enough to justify a fair degree of confidence in the model's predictions.

APPLICATION 1: NUMBERS OF LIVING CHILDREN OF SOUTH KOREAN WOMEN, 1970–2015

Figure 3 shows projections from 1990 to 2015. (The year 1990 is the most recent date for which I have been able to obtain census data). Figure 3 also shows estimates from previous censuses. The proportion of older women with two or three living children is evidently set to rise sharply over coming decades and the proportion with larger numbers of living children is set to fall. Anyone who tried to predict future numbers of living children for women aged 60–69 by simply extrapolating trends up to 1990 would have been led seriously astray. Initially, however, these changes will merely have the effect of reversing earlier upward trends in numbers of living children.

The relationship between the trends in numbers of living children and the trends in mortality and fertility can be seen in Figure 4. Cohort B, which was aged 60–69 in 1995, experienced approximately the same fertility rates during its peak child-bearing years as cohort A, which

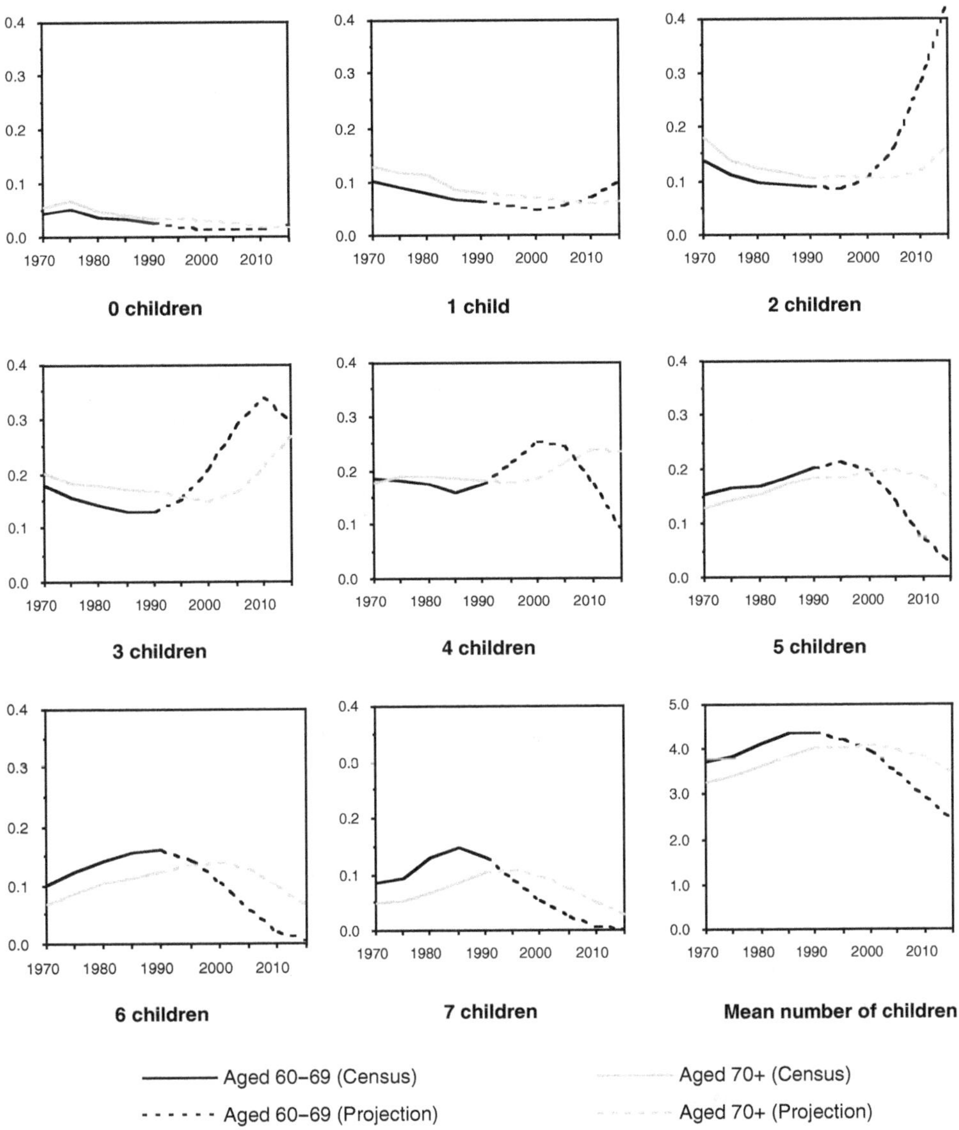

Figure 3. Proportion of South Korean women aged 60–69 and 70+ with the indicated number of living children, from censuses and projection.
Sources. 1970–1990: calculated from census data presented in UN Population Division (2000: Table 8). 1990–2015: output from projection model.

was aged 60–69 in 1970. The mortality rates faced by cohort B's children were, however, lower. As a result, cohort B entered old age with a higher number of living children than cohort A. Cohort C experienced much lower fertility than cohort B, and will enter old age with significantly fewer living children.

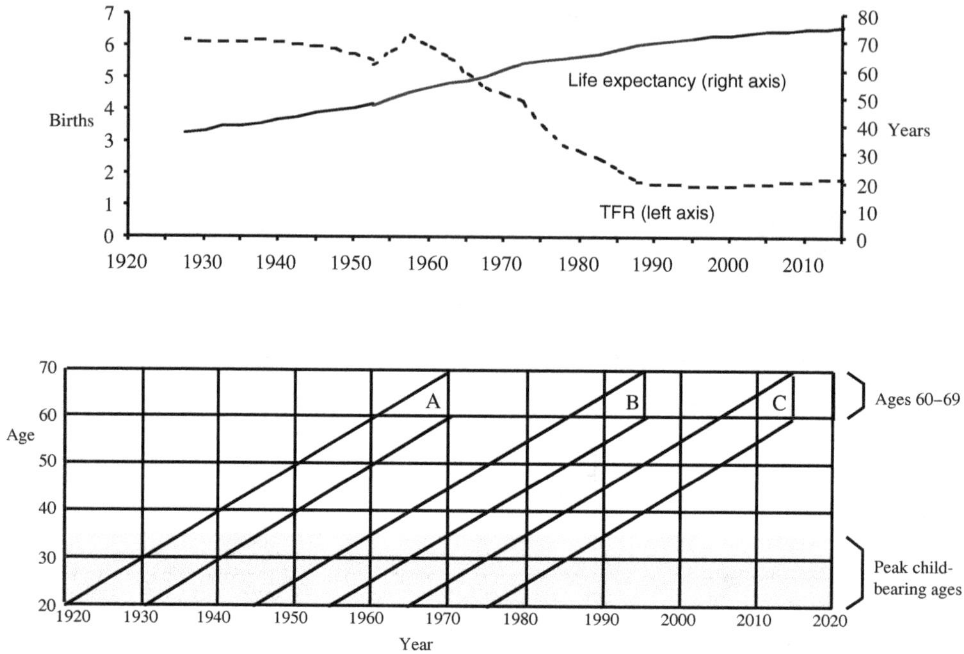

Figure 4. Demographic trends in South Korea, with Lexis diagram.
Sources: Estimates for total fertility rate and joint life expectancy at birth, 1925–1950 (Kwon 1977: Tables B1 and D4). Estimates and projections for total fertility rate and life expectancy, 1950–2015 (UN Population Division 1999).

APPLICATION 2: THE EFFECT OF THE HIV/AIDS EPIDEMIC ON NUMBERS OF LIVING CHILDREN IN THAILAND

Wachter Knodel, and VanLandingham (2002) use a combination of micro-simulation and macro-simulation to investigate losses of children to HIV/AIDS among the cohort of Thais aged 50 or above in 1995. They estimate the proportion of old people likely to eventually lose children, and the timing of the children's deaths. They conclude that around 11–13 percent of the cohort will eventually lose at least one child.

I have used the projection model to construct a somewhat different set of estimates: the distribution of Thai women aged 60 and over by numbers of living children in the years 1990–2015, with and without the AIDS epidemic. The calculations are necessarily tentative: the Thai data on numbers of living children and on AIDS mortality are not sufficiently reliable to allow any great precision. However, the results seem to be consistent with those of Wachter et al., and the main qualitative conclusion – that the effect of the epidemic on numbers of living children is moderate – seems fairly secure.

The UN Population Division's (1999) estimates and projections for Thailand take account of AIDS deaths. Subtracting the UN's (1999a) estimates of excess AIDS mortality from

Table 2. The effect of the HIV/AIDS epidemic on the distribution of Thai women aged 60+ by numbers of living children.

	Percent distribution of women by number of living children					
	(a)	(b)	(c)	(d)	(e)	(f)
Living children	1990	2015, without AIDS	2015, with AIDS	2015, with AIDs, deaths correlated	(c) as a percentage of (b)	(d) as a percentage of (c)
0	7.6	11.3	11.7	11.7	103.3	99.9
1	12.2	11.1	12.1	12.2	109.0	99.9
2	11.1	21.4	22.1	22.1	103.5	100.0
3	15.2	22.0	21.9	21.9	99.5	100.1
4	15.3	15.5	14.9	14.9	96.7	100.1
5	13.6	9.1	8.6	8.6	94.1	100.1
6	9.7	4.9	4.5	4.5	91.8	100.0
7	6.6	2.5	2.2	2.2	89.7	99.9
8	4.1	1.2	1.1	1.1	86.8	99.5
9	2.4	0.6	0.5	0.5	85.3	99.1
10+	2.3	0.5	0.4	0.4	82.5	98.6
Total	100.0	100.0	100.0	100.0	–	–
Mean	3.92	2.94	2.84	2.84	96.7	100.0

the rates incorporating AIDS deaths provides an approximate description of how the Thai population would have evolved without the epidemic. Column (c) in Table 2 shows projection numbers of children based on the with-AIDS scenario while column (b) shows results based on the without-AIDS scenario. The starting point for both is the 1990 census distribution of women by numbers of living children. Virtually all AIDS deaths in Thailand have occurred after 1990. Comparing columns (c) and (b) gives an indication of the effect of the AIDS epidemic on numbers of children.

Comparing the mean numbers of living children at the bottom of columns (b) and (c) indicates that women aged 60 or more in 2015 will, on average, have lost about 0.1 children to AIDS. This is broadly consistent with Wachter et al.'s estimate that 11–13 percent of the cohort aged 50 or over in 1995 would lose one or more children over their lifetime.

The AIDS epidemic has increased the number of women with few children and reduced the number of women with many children, compared with the expected pattern if the epidemic had not occurred. The changes are fairly significant for women with many children. These women form only a small part of the total population, however, and the overall effect of the AIDS epidemic on the distribution is not as large as might have been expected. The effect is much smaller than the ongoing effects of the demographic transition, as can be seen by comparing the 1990 distribution with either of the 2015 distributions.

Mortality rates vary between classes, occupations, regions, and other social categories. Since members of the same family tend to belong to the same categories, their chances of dying are correlated. This is certainly true for AIDS deaths, because, for example, some families have several members working as sex workers while most have none. The projection results cited so far, however, assume that all families face the same risks. As mentioned above, the program allows users to test whether modifications to this assumption affect the results significantly. Users can define a high-risk group, specifying the proportion of the population which belongs to this group and the difference between the high-risk group's mortality rates and those of the rest of the population. Column (d) of Table 2 shows the distribution in numbers of living children that result from assuming that one quarter of the Thai population have three times the mortality rates of the remaining three quarters. The difference from the no-correlation case is never larger than a tenth of a percentage point, and so does not show up when comparing columns (c) and (d). Examination of column (f) shows, however, that the existence of correlations between deaths bunches the distribution toward the middle slightly. The reason the effect is so small is that correlations between deaths show up as rises or falls in the numbers of multiple deaths in families. Since deaths are themselves rare events, multiple deaths are very rare events, even within the high-risk group.

POSSIBLE EXTENSIONS TO THE MODEL

It would be useful to be able to disaggregate living children by sex. The simplest approach to doing so would be to assume that the likelihood of any combination of sons and daughters could be calculated from the probability that a given child was a son. For instance, if a woman had three children, and the average probability of a given child being a son was 0.55, then the probability of having three daughters would be $(1 - 0.55)^3$. Unfortunately, however, the populations where interest in the number of sons and daughters is greatest are exactly the ones where this approach is likely to fail. The use of sex-selective abortions, for instance, to avoid having too many girls means that the proportion of parents having three daughters would be less than that dictated by chance. For such populations it would be necessary to start with data on numbers of children disaggregated by sex, and to project the whole distribution. This would increase the scale of the calculations considerably, but would not introduce any new principles.

The current version of the model is restricted to elderly women, because of the lack of data on male fertility. Zeng, Vaupel, and Wang (1999) have successfully modified female age-specific rates to generate results for men. The same approach could perhaps be applied here.

A third possible extension would be to try to lengthen the projection time beyond 25 years. In countries which are at or very near to the end of their demographic transitions, it may be feasible to assume that cohorts currently aged less than 35 years will have the same child-bearing patterns as the cohort currently aged 35–39. In cases where this assumption can be safely made, extra years could be added to the projection period with very little extra effort.

References

Note: Thailand's National Statistical Office does not generally put publication dates on its reports. I obtained the publication dates cited in the text by adding two years to the date of the survey or census. This seems to yield the same publication dates as those usually cited in the Thai demographic literature.

Chayovan, N., Knodel, J. and Siriwan S. (1990), Thailand's Elderly Population: A Demographic and Social Profile Based on Official Statistical Sources. Population Studies Center, University of Michigan Comparative Studies of the Elderly Research Report No. 90–2.

Jiang, L. (1994), "Parity and security: a simulation study of old-age support in rural China." *Population and Development Review*, 20(2):423–48.

Knodel, J. and Chayovan, N. (1997), "Family support and living arrangements of Thai elderly." *Asia Pacific Population Journal*, 12(4):51–68.

Kwon, T.H. (1977), *Demography of Korea: Population Change and its Components 1925–66*, Seoul National University Press: Seoul.

National Statistical Office. (1982), *1980 Population and Housing Census*, Office of the Prime Minister: Bangkok.

National Statistical Office. (1986), *The 1984 Survey of Population Change*, Office of the Prime Minister: Bangkok.

National Statistical Office. (1991), *The 1991 Survey of Population Change*, Office of the Prime Minister: Bangkok.

National Statistical Office. (1992), *1992 Population and Housing Census*, Office of the Prime Minister: Bangkok.

National Statistical Office. (1997), *Report on Population Characteristics: The 1995–1996 Survey of Population Change*, Office of the Prime Minister: Bangkok.

United Nations Population Division. (1997), *World Fertility Patterns 1997*, New York, Wall chart.

United Nations Population Division. (1999), *World Population Prospects: The 1998 Revision*, New York.

United Nations Population Division. (1999), *The Demographic Impact of HIV/AIDS: Report on the Technical Meeting*, New York, 10 Nov. 1998, (E-SA/P/WP.152).

United Nations Statistics Division. (2000), *Demographic Yearbook Historical Supplement, 1948–1997*, ST/ESA/STAT/SER.R/CD/28 (CD-ROM).

Wachter, K.W., Knodel, J., and VanLandingham, M. (2002), "AIDS and the elderly of Thailand: Projecting familial impacts" Demography 39(1):25–41.

Zeng, Y. (1986), "Changes in family structure in China: A simulation study." *Population and Development Review*, 12(4):675–703.

Zeng, Y., Vaupel, J.W. and Wang Z. (1999), "Household projection using conventional data"' In: *Frontiers of Population Forecasting. Supplement to Population and Development Review*, Vol. 24. W. Lutz, J.W. Vaupel and D.A. Ahlburg, eds. New York: The Population Council.

CHAPTER 7. TOWARD A CONCEPT OF POPULATION BALANCE CONSIDERING AGE-STRUCTURE, HUMAN CAPITAL, AND INTERGENERATIONAL EQUITY

WOLFGANG LUTZ
World Population Project
IIASA, Laxenburg, Austria
and
Vienna Institute of Demography

WARREN SANDERSON
World Population Project
IIASA, Laxenburg, Austria
and
Departments of Economics and History
State University of New York at Stony Brook

1. Introduction

The communication between international population analysis and politics is in disarray these days. Before the Cairo Summit on Population and Development in 1994, the relationship between economic growth, social development, and environmental change, on the one hand, and population growth, on the other, was viewed from a macro perspective. The Cairo Conference, in the view of many observers, essentially transformed population from a development concern to a human rights issue. Recently, ageing has been added to the agenda leaving us with three questions, but no integrative framework within which to answer them. In this paper we provide the first steps towards developing such a framework. We call it population balance.

Population balance considers the role of the population variable in development and environmental change, and its role with respect to intergenerational equity. It transcends the conventional focus on total growth rates and looks at age-specific growth rates and other age-structure features. This emphasis on age structure is consistent with the recent economic literature that demonstrates that age-structure effects are important in explaining observed variation in economic growth rates (Bloom, Canning, and Malaney 1999; and Bloom and Williamson 1998, for example). It also transcends the realm of strictly demographic variables and incorporates a key source of population heterogeneity, namely, the level of education.

S. Tuljapurkar et al. (eds.), Population, Resources and Development, 119–137.

Before we go into the political dimension of this concept, and later into the analytical and empirical aspects, it is worth presenting a few thoughts on the way the words population and balance are used here. Population is used in its original macro-level meaning describing a well-defined group of human beings with emphasis on quantitative aspects, which are primarily its total size and growth rate, but also on its changing distribution by age and sex and place of residence. There are other population-based characteristics, such as marital status, labour-force participation, ethnicity, and educational status, which do not represent core demographic dimensions, but often enrich population analysis in a meaningful way. To such sub-populations the demographic tools of multi-state cohort-component analysis and projection can be readily applied. In the context of population balance, educational attainment is considered an additional population dimension because of its paramount importance for both the determinants and consequences of population change. Education is not only one of the most important sources of population heterogeneity (Lutz et al. 1999) but can also be readily measured and projected for all parts of the world (see Lutz and Goujon 2001).

The word balance does not have any well-defined meaning in population studies. In economics it is sometimes used in the sense of a dynamic consistency of certain elements of economic structure. In this context, balance has a highly dynamic meaning, more like 'balancing'. Imagine a person walking on a log in the water being exposed to all kinds of currents and winds and trying not to bend too far to either side and fall into the water. To stay in balance the person on the log must always be moving. In the real world, population policies must keep adapting if the population is to stay in balance.

If, in a political sense, there is a potential model for the notion of population balance then it is the notion of 'sustainable development'. Although it is very poorly defined and people interpret it in surprisingly different ways, it still has become, politically, an extremely powerful notion; the whole UN system is currently being restructured under the notion of sustainable development. While we hope to define population balance in a more precise and operational way than 'sustainable development', it is also worth noting that the notion may benefit from not being too narrowly defined. Different people concerned with different aspects of the issue can read different things into it. It could mean the balance between population and the environment; it could mean a balance in the sex and age distribution, a balance across generations, etc. Hence, the challenge is to define the notion widely enough to make it relevant for different groups, but at the same time to provide a choice of different precise definitions and an analytical toolbox that allows for rigorous scientific analysis.

There is a second way in which the concept of population balance is related to the notion of sustainable development. Both have at their heart comparisons of welfare across generations. Imagine a country undergoing a demographic transition with a government planning a dynamic educational strategy. Some strategies could cause imbalance in intergenerational welfare with some cohorts paying much more for the educational burden than they receive in benefit, while others benefit far more than they pay. Population balance can help guide policy-makers toward dynamic educational strategies where burdens and benefits are shared more equitably across generations. Governments do not control steady-state rates of population growth, but in a purely theoretical analysis we can imagine a 'central planner' who does. Population balance can show the 'central planner' where rates of population growth are so high or so low that generational welfare suffers. In between, there is a wide range in which population can be said to be in balance.

The strength of the population balance concept is not so much that it can perform the analysis of ageing on the viability of pension plans or the analysis of the effects of population growth on economic growth better than stand-alone models. Its strength is that it shows how cohort welfare, macroeconomic performance, educational levels and ageing are all related through the application of age-structure dynamics.

In section 2, we discuss the need for the concept of population balance. In section 3, we use a new measure called the Cohort Succession Ratio (CSR) to visualize the dimensions of age-structure change in real populations. Simulations of a highly simplified population balance model appear in section 4. We consider balance in terms of different total fertility rates and different educational levels in stable populations, and in terms of fertility and education-transition dynamics. Section 5 concludes with some thoughts on how we plan to develop the population balance framework and how it relates to other papers in this volume.

2. The International Political Dimension: Trying to Link Concerns about Population Growth, Ageing, Education and the Relationship between Generations

After decades of much concern and concerted action to deal with the population explosion, recent news about the end of world population growth to be expected during the second half of this century (Lutz et al. 2001), and fertility declines that are more rapid than expected, seem to make these concerns less relevant, even though in some world regions very strong growth is still expected. In parallel, the process leading up to the 1994 Cairo Conference on Population and Development has brought a strong shift of emphasis from macro-level population concerns to individual rights and reproductive health concerns, i.e., an almost exclusive focus on the individual welfare dimension. Macro-level population-related considerations, which also impact on individual welfare, but in a more indirect way, were not given much attention. As an institutional consequence, for example, in many national development agencies the population offices, if they still exist at all, have become subsections of departments of health.

The population issue has largely become a specific sub-concern of health. But there is clearly more to population than one aspect of human health, the age-structural aspects discussed in this book being a good example, population-environment interactions being another important one. In the famous 'Cairo Consensus', the Cairo Programme of Action mentions some of these macro issues, but there is no convincing substantive link between the main focus on reproductive health and the macro-level consequences of fertility trends. One cannot help feeling that these macro concerns were simply added in an unconnected fashion in order to satisfy some of the people who "were still stuck in the population concerns of yesterday," as some delegates put it informally. Reproductive health and individual welfare more generally are clearly very important aspects of population, but there is an urgent need to link them substantively to the population related macro-level concerns.

Simultaneously, but completely unrelated to this 'Cairo Concensus', concern about population ageing grew in some of the developed countries, especially in those that have been experiencing fertility well below replacement for several decades. This macro-level population concern tends to concentrate mainly on the sustainability of existing pension schemes, but also relates to expected impacts on productivity and international competitiveness and more recently on the need to rethink immigration policies. There is a rapidly increasing scientific literature and a rapidly increasing international concern. Yet in international fora the ageing concerns tend to be treated in a completely unrelated way to the population growth and population-environment concerns and the individual-level reproductive health, female empowerment and freedom of choice concerns. There seems to be a need for some sort of an umbrella concept that links these two macro-level population concerns to each other and relates them to these individual-level concerns.

Another issue that has gained prominence over the last two decades is that of intergenerational equity. Interestingly, this concern originally arose independently from all of the above-mentioned population discussions although it is now clearly related to the concern about population ageing. But the intergenerational dimension came to international political prominence through environmental concerns and the very definition of sustainable development (as given by the Brundlandt Commission – the World Commission on Environment and Development 1987). This definition is explicitly based on intergenerational concerns, i.e., improving living conditions without compromising the possibilities of future generations to meet their needs. More recently in the context of pension reforms in Europe that will become necessary due to expected population ageing and the great dependence of pay-as-you-go pension systems on relative sizes of age groups, there is increasing concern over the fact that an ever-increasing burden of the pension load is put on the shoulders of the younger and smaller cohorts. The degree to which the younger generations will be unduly hit by this depends, of course, also on the future course of productivity gains. This second factor will be approximated by the concept of education in this paper.

Education and human capital formation have become another important force seen to influence both the consequences of rapid growth and rapid ageing, because they are directly related to the productivity changes associated with changes in the age composition. Especially in the context of developing countries the World Bank has recently announced that education would be its top priority. But future productivity gains that are likely to be associated with higher technical skills are also seen as a possible solution to the problem of ameliorating the consequences of population ageing. Since the educational composition of the total population and the educational enrolment rates of the young cohorts are not independent from age-structural changes (e.g., if the youngest cohorts grow too rapidly, enrolment rates cannot keep pace) these two things should be viewed together. Hence both age structure and education, independently and in their interaction, play an important role in economic growth and consequently in intergenerational equity. The joint consideration of both may serve as a binding element between growth and ageing concerns under a concept of population balance.

3. The Dynamics of Changing Age Structures: Analytical Approaches and Empirical Applications

When talking about analytical approaches to describe and analyze age-distributional changes over time, fairly simultaneous but independent work by Sam Preston (Preston and Coale 1982; Horiuchi and Preston 1988) and Nathan Keyfitz (1987, 1990) comes to mind. In the following we will briefly discuss their approaches and give some empirical illustrations. We will also propose a new indicator tentatively called 'cohort succession ratio' (CSR).

Following earlier work by Ansley Coale on a generalization of the stable population model to deal with variable growth rates, Horiuchi and Preston (1988) published a paper in demography entitled "Age-Specific Growth Rates: The Legacy of Past Population Dynamics." This paper demonstrates how the widely available but rarely used sets of age-specific growth rates in a population can be used to reconstruct all the pertinent features of a population's demographic history that are required to relate major demographic functions for a particular period to one another. They derive a formula,

$$\ln \frac{N(a,t)}{n(a,t)} = - \int_0^a r(x,t)\, dx,$$

which implies that the current age distribution $N(a,t)$ can be obtained by deriving the underlying stationary distribution $n(a,t)$ and converting it into the actual distribution by using age-specific growth rates, $r(x,t)$, at age x and time t.

The density function of the stationary population is derived from current births and mortality schedules as given by

$$n(a,t) = B(t)p(a,t),$$

where $n(a,t)$ is the size that the cohort aged a at time t would attain if the cohort had the size $B(t)$ at birth and had mortality experiences as observed at time t.

This concept has been used mainly to make estimates from incomplete data. But age-specific growth rates also provide useful information about the history of a population for periods during which vital rates were not recorded. Whether and how this approach can be used in addressing future population-related challenges needs further exploration.

Nathan Keyfitz's work on the issue essentially grew out of his empirical work in advising the Indonesian government on population-related challenges and the analysis of past and likely future world population trends. It may be best known under the keyword "international youth cohort" (Keyfitz 1990). This work focused both on the reconstruction of the onset of the post WWII population growth and on the possible consequences of age-structural discontinuities. In his 1987 paper entitled "The Demographic Discontinuity of the 1940s" he reconstructs the onset of the "population explosion" through the analysis of intercohort increases as given in more recent censuses. As an analytical tool Keyfitz mainly uses difference analysis (Keyfitz 1990). By describing the first and second differences between the sizes of subsequent cohorts in an often graphical manner, the demographic discontinuities become readily visible.

Although much of Keyfitz's work in this field is being used for the same purpose as that of Preston and others described above, namely the reconstruction and decomposition of past trends, Keyitz also extended his work into the future and made projections of likely future discontinuities as embedded in the pattern of cohorts already born. Occasionally, he drew inferences beyond the limits of demography, e.g., by suggesting that during the mid-1990s Indonesia may see some major political shake up due to the numerous rather well-educated youth cohort entering the labour market (Keyfitz 1988). This bold forecast turned out to be quite accurate.

In studying intercohort changes, whether it is done through age-specific growth rates or differences between cohort sizes, both Preston and Keyfitz limit their focus to the comparison of adjacent cohorts. In the broader context of studying the implications of population ageing and issues of intergenerational equity, it is useful to apply a more general scheme that generally describes the relationship between any two cohorts at a given time. To do this we introduce the notion of cohort succession ratios.

A cohort succession ratio can be defined as follows,

$$CSR(a,x,t) = N(a,t)/N(a+x,t)$$

with N standing for total population (of both sexes) and the indices a for age (indicating either single or five-year age groups), t for time, and x for the age difference in years for the cohorts to be compared.

In the case of $x = 1$ (or 5 in the case of five-year age groups) the pattern of CSRs should be very similar to that of age-specific growth rates and first differences because all compare adjacent cohorts with only the metric being different.

Figures 1a–e give empirical illustrations of the patterns of the different measures discussed over age and time for China. The data are five-year age groups from 1950 to 2050 as given by the UN (1999) estimates and projections. Figure 1a simply gives a 3-D plot of the total population of China (without Hongkong) for men and women combined. It shows that in 1950 China had a very young age distribution with some scars of history. Up to the 1970s the population at all age groups (interrupted by the turmoil of the 1960s) grew sharply with the movement along cohort lines clearly visible as ridges on the graph. Mortality improvements are shown by the fact that the ridges become more horizontal over time, i.e., there is less attrition of cohorts over time. After that low fertility results in declining initial sizes of birth cohorts with strong echo effects from the adult age distribution.

The first differences (calculated as the differences in size between the same five-year age groups in year $t+5$ minus that in year t) are plotted in Figure 1b. They give an even more pronounced picture of the consequences of the fertility trends and the associated echo effects on the starting sizes of cohorts. After the period of high fertility in the 1950s, there was a short but steep cut around 1960 associated with the great leap forward (which with some imagination is also seen in terms of higher mortality through a different inclination of the ridge). After recovery of fertility, another steep decline followed during the late 1960s. Finally, during the 1970s the consequences of the strict one child per family law became evident. The picture is somewhat distorted due to the fact that it only looks at the trends of five-year age groups. The data base probably includes a lot of estimated and smoothed data, especially for the older cohorts. The graph of second differences (Figure 1c) shows the same pattern in an even more pronounced fashion.

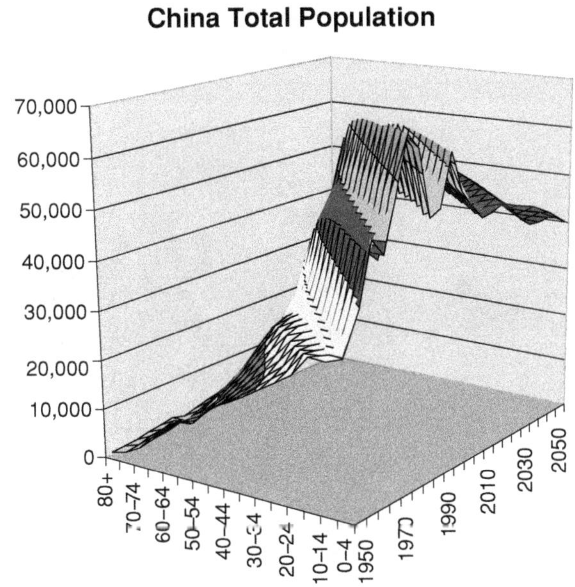

Figure 1a.

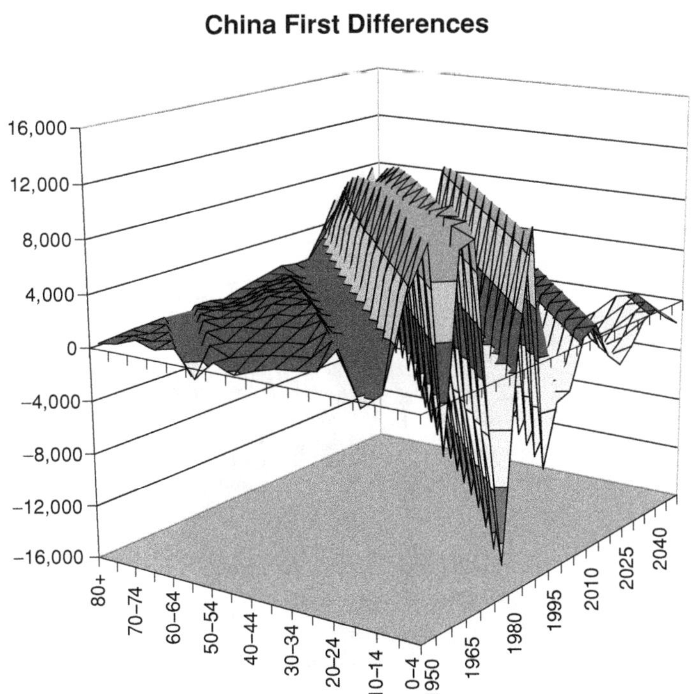

Figure 1b.

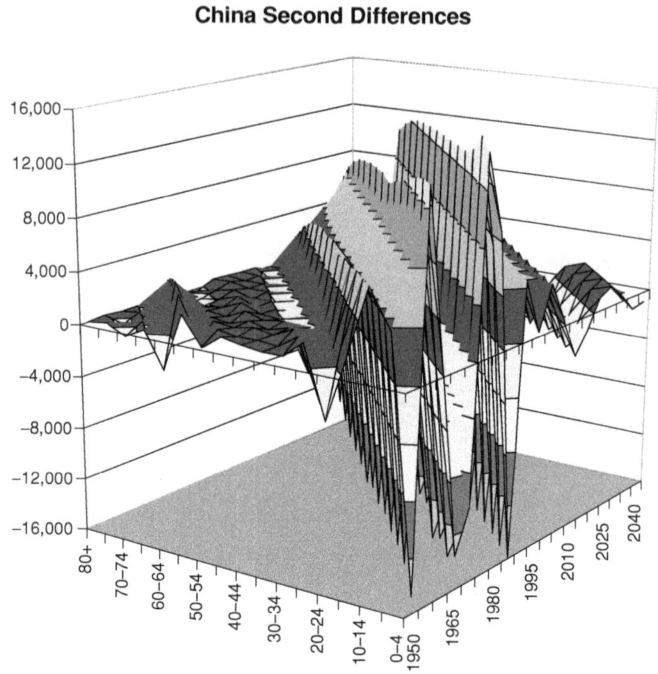

Figure 1c.

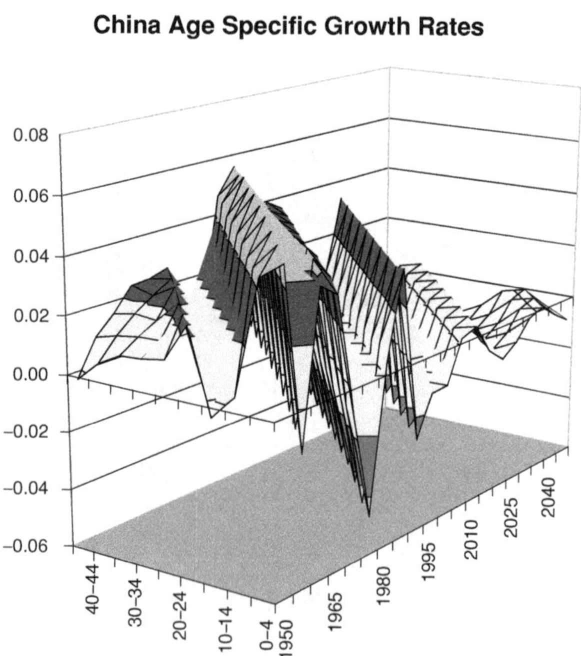

Figure 1d.

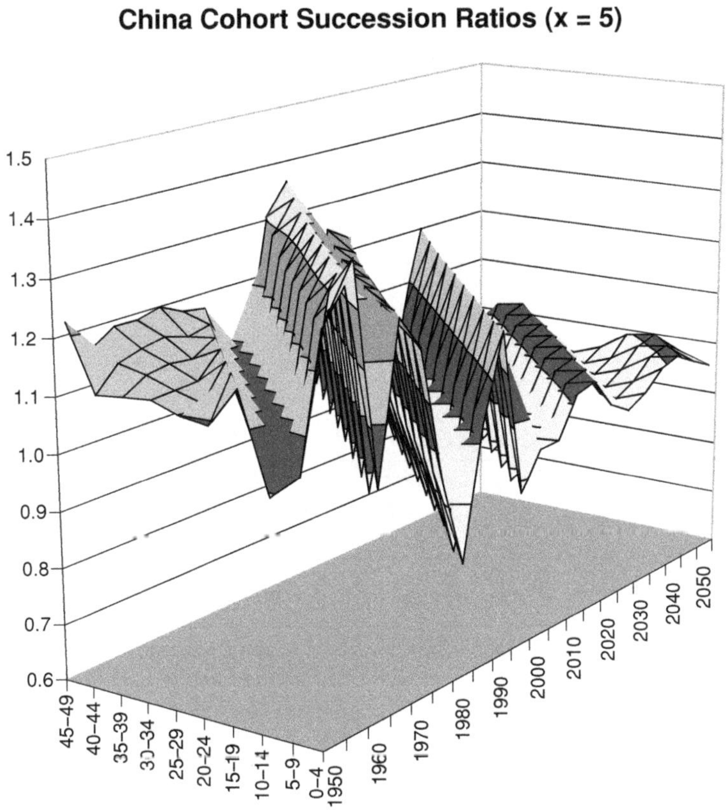

Figure 1e.

Figures 1d and 1e give the age-specific growth rates and the cohort succession ratios for subsequent cohorts. As expected these two indicators show rather similar patterns. Since these measures operate on a relative scale and are more sensitive to data problems, especially in the older age groups with few members, these graphs are restricted to the population below age 50. One thing is very clear from the comparison of all these figures. The strong fluctuations in Chinese fertility during the past few decades are likely to produce strong waves of echoes, even in the case of almost constant future fertility as assumed by the UN.

One strength of the CSRs, as compared to the other measures, is the possibility of comparing non-adjacent cohorts. Figures 2 to 4 present such indicators for China, Thailand and Singapore. One ratio compares the youngest cohort (age 0–4) to the cohort aged 25–29. Since this 25-year difference is close to the mean age of childbearing, the ratios can be roughly interpreted in the sense of a gross reproduction rate that to some degree adjusts for infant and child mortality, or a net reproduction rate not considering the mortality of ages 5–24. Before 1970 this ratio tends to be above two in all three countries, indicating that the size of the youngest age group was more than twice that of the age group 25–29. In all three countries fertility decline brings down the ratio thereafter. In Singapore, a level

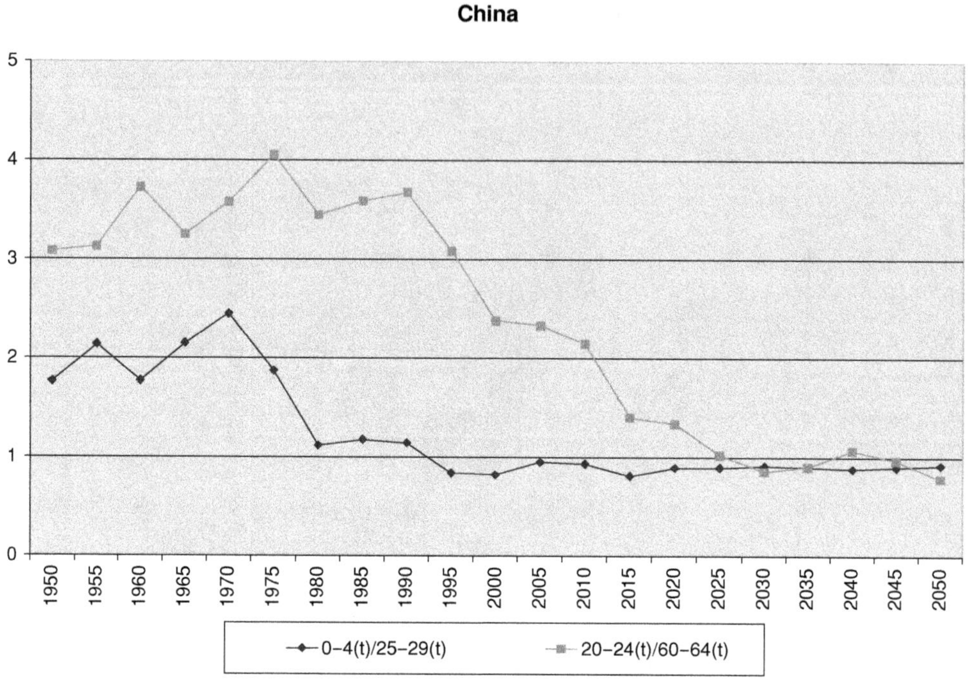

Figure 2.

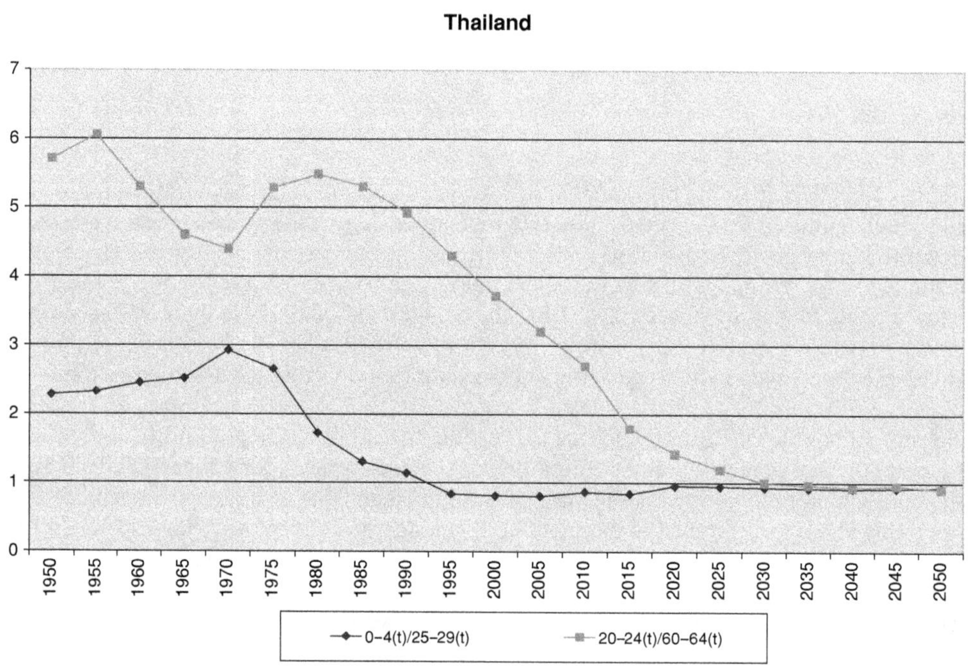

Figure 3.

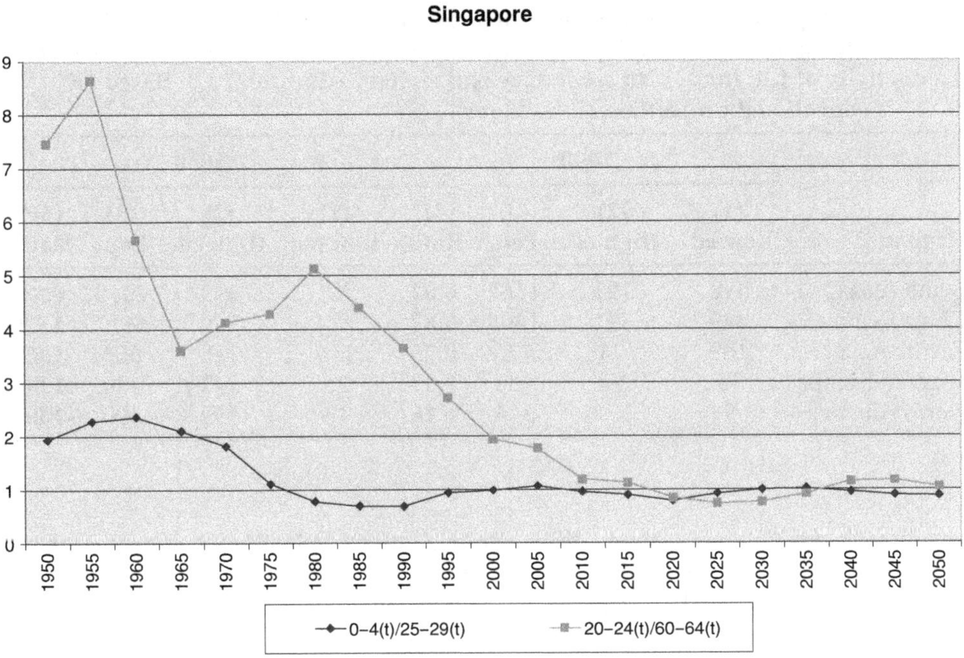

Figure 4.

below 1.0 was already reached in 1980. Singapore is special in the sense that immigration of young adults has a great influence on these ratios.

The other ratio plotted in Figures 2 to 4 is of those aged 20–24 (a typical age for entering the formal labour market) and those aged 60–64 (a typical age for retiring, at least in industrialized countries). Hence, as a rough approximation this ratio could possibly describe labour-force replacement (those entering divided by those leaving). If this ratio is above 1.0, the labour force is growing. In all three countries we see a very strong growth in the labour force between 1950 and the present. As to the future, Singapore will first reach a situation in which the labour-force replacement will fall below 1.0. In Thailand and China, this is only likely to happen after 2020.

We have to do more to move this analysis toward incorporating the aspect of human capital formation. This is important for answering the question whether, e.g., a labour-force replacement ratio below 1.0 is something to be concerned about. Since educational attainment of the labour force is a most important determinant of future productivity, and hence impacts on the future standard of living and intergenerational equity, we will have to find a way to include it into these calculations.

Table 1 gives some very simple calculations based on the recently produced projections of educational attainment for 13 world regions (Lutz and Goujon 2001). These first global forecasts of the population by educational attainment distinguish between four educational groups: no education; some primary; some secondary; and, some tertiary. For the purpose

Table 1. Change in age- and education-dependent productivity for selected world regions. Men and women aged 20–65 with no and some primary education have a productivity of 1.0; those with secondary and tertiary education, 2.0. Based on IIASA's educational projections (in millions).

	2000				2030			
Region	(1) Low ed.	(2) High ed.	(3) Pop.	(4)* Ratio	(1) Low ed.	(2) High ed.	(3) Pop.	(4)* Ratio
South Asia	498	191	1367	**0.64**	731	492	2030	**0.84**
China Region	449	391	1408	**0.87**	274	749	1617	**1.10**
Pacific Asia	169	94	476	**0.75**	179	218	605	**1.02**
Western Europe	75	201	456	**1.05**	25	247	480	**1.08**
North America	9	177	314	**1.16**	15	199	390	**1.06**

*(4) = [(1) + 2 × (2)]/(3)

of simple calculation, we combine them into two groups and simply assume that those aged 20–65 with no or some primary education have a productivity of 1.0, and those aged 20–65 with some secondary or higher education have a productivity of 2.0. Those below age 20 and above age 65 are assumed to be not productive in this model. Table 1 presents a ratio calculated in the following way: the number of working age (20–65) men and women with higher education is multiplied by the factor of 2.0 and added to the working-age population with lower education. This sum is then divided by the total population (all ages) in the given year. This ratio can be interpreted as a measure of the age- and education-dependent productivity of given populations. The factor 2.0 is used here to make the numbers in Table 1 consistent with the model in section 3.

Table 1 presents these calculations for five selected world regions. South Asia has the lowest ratio in 2000 and is likely to have the lowest one in 2030, despite some improvements. The main reason for this poor performance is the low educational level of the total population combined with still high youth dependency. The China region, which has a comparable total population size in 2000, has much better starting conditions in 2000 (even better than South Asia will have in 2030) due to strong past investments in education and lower fertility. By 2030 China is likely to reach a high level of 1.10 which, according to our specifications, will be higher than the age and education-dependent productivity ratios for Western Europe and North America (of course, these measures do not give special attention to highly-advanced technological skills). But these trends clearly indicate that China will strongly gain in terms of international competitiveness. The region of Pacific Asia, which includes most of Southeast Asia, is also set for an impressive improvement. While being between China and South Asia in 2000, it is likely to have even more impressive gains in age- and education-dependent productivity than China.

But even this weighting by education does not yet bring us to a full operationalization of intergenerational equity concerns in the context of population balance. To make further progress in this direction, for the time being we will leave the realm of empirical data and enter the world of simulation.

4. A First Attempt to Simulate Age, Cohort and Human Capital Aspects of Population Balance

Population balance can mean many things. For example, it could mean population in balance with its environment. It could mean a balance of sex or age groups. It could mean a population whose growth is in balance with the growth of the capital stock and infrastructure that it needs for its production. But it would be difficult to understand population balance if we were to integrate them all into a single very complex framework. Instead, we go to the other extreme of producing a simple model of population in balance between its period aspect (production) and its cohort aspect (lifetime consumption). By studying first this aspect of population balance and later adding other elements to it, we hope eventually to produce a rich understanding of population balance.

Population in balance with its environment is commonly discussed under the rubric of 'sustainable development'. In that context, the definition of balance is clear. Generations that come first are not to consume and produce in such a way that future generations are denied the possibilities that the earlier generations possessed. More briefly, balance is intergenerational equity. Because of the importance of intergenerational equity itself and because we anticipate soon adding environmental balance to our framework, we interpret balance here to mean intergenerational equity.

We assume a one-sex population closed to migration and with fixed age-specific mortality rates. There are two types of people: those with education and those without education. Two things are allowed to vary exogenously over time: age-specific fertility rates, summarized by the total fertility rate (TFR), and probability of a young person becoming educated.

The life cycle is assumed to have three stages. In the first, children consume, either acquire education or not, but do not produce. In the second, adults consume and produce both children and output. In the third, retirees consume, but again do not produce output. Output in any year depends on the numbers of educated and uneducated people in the second stage weighted by their relative productivities.

$$Y(t) = N(2,t,u) \cdot \pi(u) + N(2,t,e) \cdot \pi(e), \tag{1}$$

where $N(2,t,u)$ is the number of uneducated people in life cycle stage 2 in year t, $\pi(u)$ is the productivity of unskilled workers, and $N(2,t,e)$ and $\pi(e)$ have the analogous interpretations for educated workers.

People aged 6 and above in stage 1 can obtain one year of education every year. Once enrolled in education at age 6, they remain in school or other training to the end of the stage. The enrolment rate at age 6 changes exogenously over time. Each year of education is assumed to cost E euros per year. The total cost of education in year t is simply the product of the education cost and the number of children in school.

$$EDCOST(t) = N(1,t,e) \cdot E \tag{2}$$

where $N(1,t,e)$ is the number of people in stage 1 who are in school and E is the per annum cost of schooling.

Output is divided among the people alive at time t. People in stage 1 get to consume a fraction, $w(1)$ of the amount allocated to people in stage 2. People in stage 3 get to consume a fraction $w(3)$ of the amount allocated to people in stage 2. The two fractions are considered exogenous here.

The income of a person in stage 2 in year t, then, is just $I(2, t)$, where

$$I(2,t) = \frac{Y(t) - EDCOST(t)}{w(1)\cdot(N(1,t,u)+N(1,t,e)) + (N(2,t,u)+N(2,t,e)) + w(3)\cdot(N(3,t,u)+N(3,t,e))}. \tag{3a}$$

The income of a person in stage 1 in year t is

$$I(1,t) = w(1)\cdot I(2,t), \text{ and} \tag{3b}$$

$$I(3,t) = w(3)\cdot I(2,t) \tag{3c}$$

The lifetime average income of a person born in year c who survives to his/her 100th birthday is:

$$LIFE(c) = \left[\prod_{a=0}^{a1} I(1,a+c)\cdot \prod_{a=a1+1}^{a2} I(2,a+c)\cdot \prod_{a=a2+1}^{99} I(3,a+c)\right]^{1/100}. \tag{4}$$

Before we look at simulation results from this very simple model, it is perhaps important to strongly emphasize that we do not interpret this to be a realistic model of population and production. To take the model in the direction of greater realism, we would have to incorporate factors such as the capital stock so that we could capture the capital dilution effect of faster economic growth; the environment, so that we could capture diminishing returns generated by fixed resources; and, endogenous technological change, so that we could incorporate the effects of scale and education on the rates of invention and adoption of new technologies. All of these would greatly complicate the model and distract us from the main point that we wish to make here. In the simplest imaginable age-structured model of population, which takes into account both its period aspect (production) and its cohort aspect (lifetime consumption), balance is best.

This model differs from standard economic models by having no discount rates. The logic of balance in the sense of sustainable development requires that all cohorts be treated equally; this rules out discounting across cohorts. Discounting within cohorts is a different matter. Instead of Equation (4), we could have chosen to look at expected lifetime income, where the expectation was taken over the probabilities of surviving to different ages. Given that we used a life expectancy of birth of 70 years and that it does not vary over the simulations, the results would not have been much different. We have

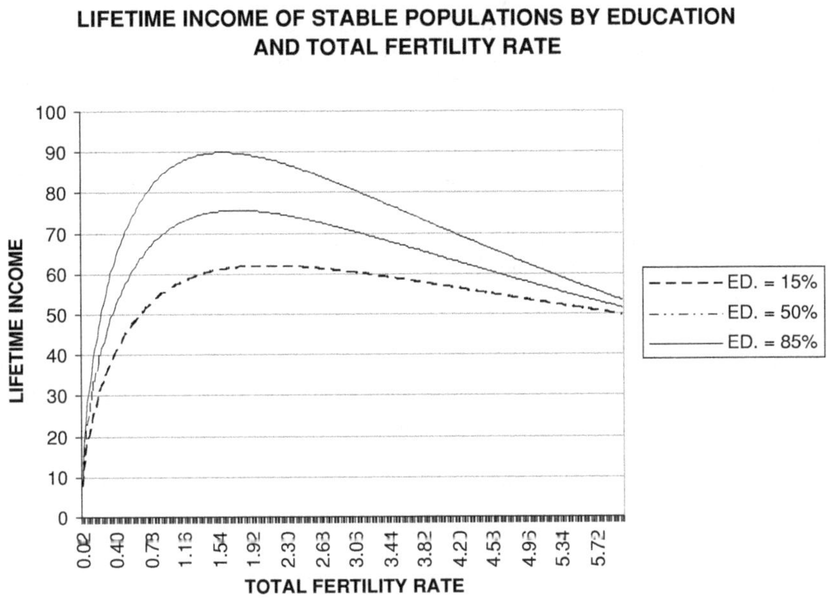

Figure 5.

tried here to be as simple and transparent as possible and not introduce any unnecessary complications.

Figures 5 through 9 show simulations of the simple model under the following assumptions:

1. Stage 1 lasts from age 0 through age 19. Education and training lasts from age 6 through age 19. Stage 2 lasts from age 20 through age 64, and stage 3 lasts from age 65 through 99.
2. Simulations were run for 200 years, resulting in complete lifetime observations on 100 cohorts.
3. All simulations are begun from stable population age distributions.
4. $\pi(u) = 100$ and $\pi(e) = 200$. Data and projections of regional populations by education using the assumption that more educated workers are twice as productive as less educated workers can be found in Table 1.
5. $E = 100$.
6. $w(1) = 0.4$ and $w(3) = 0.9$.
7. Age-specific mortality rates are taken from the Coale-Demeny model, West life table for females with life expectancy at birth equal to 70.
8. The pattern of age-specific fertility rates is that given by the U.S. Bureau of the Census for Argentina in 2000.

Preliminary sensitivity tests show that the results concerning population balance that we see below are robust to plausible changes in these assumptions.

Figure 5 shows the lifetime incomes of stable populations for three levels of proportion educated for total fertility rates ranging from 0.01 to 6.0. Given the mortality rates that were used, replacement fertility occurs at a total fertility rate of 2.13 (for ease of comprehension, total fertility rates are transformed into their usual metric for a two-sex population). The first thing to notice about this graph is that lifetime income in stable populations has a single interior peak in the range of relevant total fertility rates. Total fertility rates, and therefore population growth rates, that are too high lead to lower average lifetime incomes. Total fertility rates that are too low, and therefore lead to rapid population shrinkage, also cause lower average lifetime incomes.

The shapes of the curves in Figure 5 are the result of two opposing forces. With the mortality rates used here, higher total fertility rates cause the fraction of the stable population in stage 2 to decline. This, along with the increase in the population in stage 1, causes the numerator of Equation (3a) to decrease. This is offset to some extent by the age structure effects at stages 1 and 3. According to our assumptions, the population in stage 1 has a lower command over income than the population in stage 3. Higher total fertility rates, therefore, also cause the denominator of Eq. (3a) to decrease.

Figure 5 also shows that the peak level of lifetime income occurs at lower total fertility rates for more educated than for less educated populations. This happens because educating children is a costly process that must occur prior to recouping its benefits. Age structures with lower proportions in stage 1 are therefore somewhat more advantageous.

In Figure 6 we have plotted average lifetime incomes in stable populations for three levels of fertility and for proportions educated between zero and one. Regardless of the level of education, lifetime income is always higher when the total fertility rate is at replacement than when it is either 1.0 or 3.0. This is just another way of seeing the advantage of population balance.

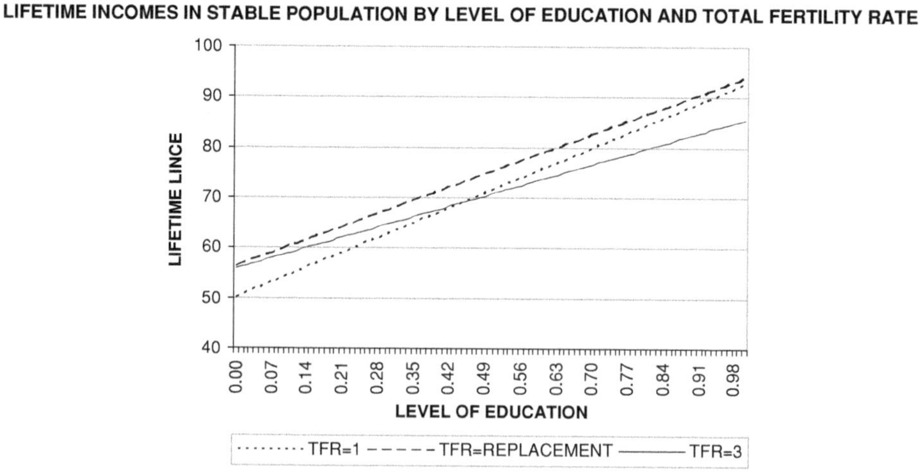

Figure 6.

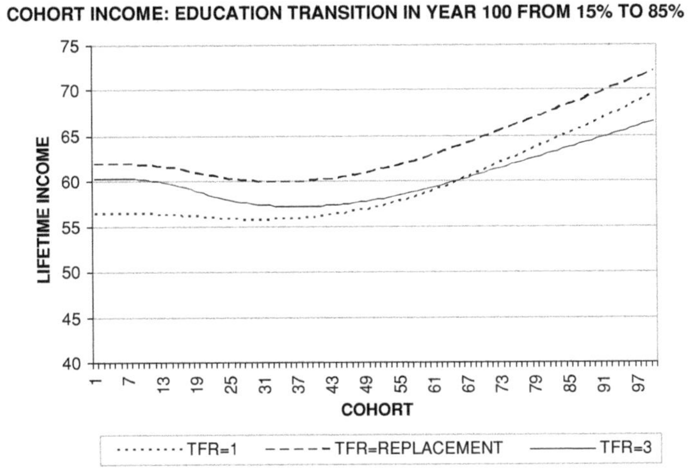

Figure 7.

In Figure 7 we can see the effects of an education transition on the lifetime income of people in three different stable populations. The figure shows the average lifetime incomes of the 100 cohorts for which we have a complete 100 years of observation, i.e., simulation (cohort 1 lives from year 1 through year 100; cohort 2 lives from year 2 to year 101, and so on). The education transition begins in year 100, when 85 percent of the children at age 6 enter school instead of the earlier 15 percent. The three cases are again a stationary population associated with replacement level fertility, a growing population with a TFR of 3.0 and a shrinking population with a TFR of 1.0.

The curves in Figure 7 fall and then rise. They fall because people in the early cohorts pay for part of the increased schooling costs when they are elderly and do not live long enough to enjoy any of the increased productivity of the educated children. It rises because eventually the increased productivity of the educated dominates the earlier educational investment costs. Again, we see that balance is best. Lifetime incomes are always higher in the population with replacement level fertility.

Figure 8 is like Figure 7 except that here we look at the effects of a fertility transition from a TFR of 4.0 to a TFR of 2.0 occurring in year 100 in the context of three different levels of education. As we would expect from what we have seen before, these fertility transitions increase average lifetime income at each level of education.

Figure 9 combines the situations of Figures 7 and 8 and looks at the joint fertility and education transitions. All simulations between with the stable age distribution are associated with a TFR of 4.0. We consider three cases. In one the education transition (an increase in the percent educated from 15 percent to 85 percent) precedes the fertility decrease (a decrease in the TFR from 4.0 to 2.0). The education transition begins in year 75 and the fertility transition begins 25 years later. In the second case, the timing is reversed and the

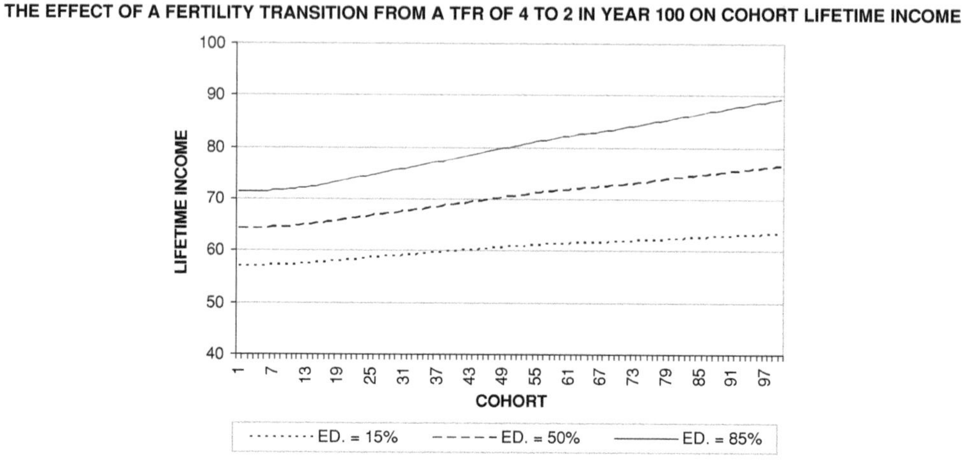

Figure 8.

fertility transition begins in year 75 and the education transition follows 25 years later. In the third case, the transitions are concurrent, both beginning in year 88.

In the case where the education transition leads the fertility transition, lifetime income of the earlier cohorts is suppressed, and lifetime income of the later cohorts is enhanced relative to the situation where the timing is reversed. If we were to consider more cohorts, however, eventually the three lines would coincide as the stable age structures become identical. We could have produced the analogs of Figures 8 and 9 for transitions from a TFR of 2.0 to a TFR of 1.0. Regardless of the timing of the transitions, lifetime incomes would fall.

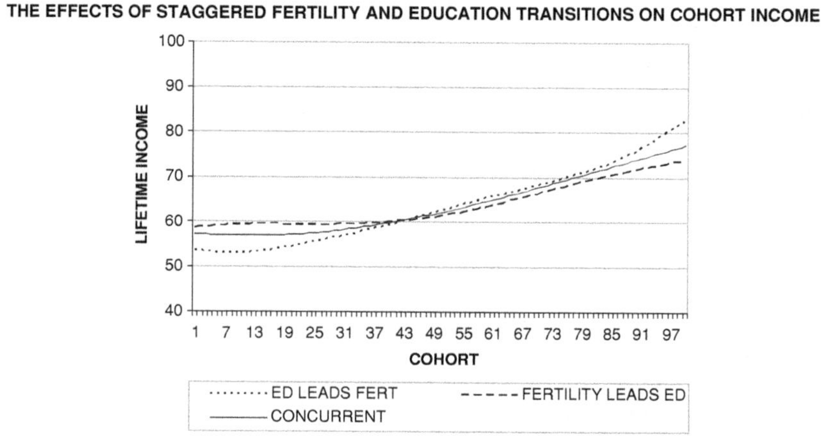

Figure 9.

We have shown that population balance appears in the simplest possible age-structured model of consumption and production. The problems of rapid population growth and rapid population ageing are indeed two sides of the same coin.

5. Outlook

This paper has presented a first and rather scattered attempt to define a new concept tentatively called population balance. It first discusses the desirability of such a concept bridging the population growth and population ageing concerns by giving explicit consideration to the implications of age-structural changes and by including human capital formation. In this paper the key criterion of sustainability, namely intergenerational equity, is discussed and simulated in the context of comparing lifetime income across cohorts. A next step will be to expand the model to explicitly incorporate environmental constraints and feedbacks. Further developments of the ideas presented here are given by Lutz et al. (2004).

References

Bloom, D.E., Canning, D. and Malaney, P.N. (1999), "Demographic change and economic growth in Asia." *CID Working Paper*, No. 15. Harvard University, Center for International Development: Cambridge, MA.

Bloom, D.E. and Williamson, J.G. (1998), "Demographic transitions and economic miracles in emerging Asia." *World Bank Economic Review*, 12:419–456.

Horiuchi, S. and Preston, S.H. (1988), "Age-specific growth rates: the legacy of past population dynamics." *Demography*, 25(3):429–441.

Keyfitz, N. (1987), "The demographic discontinuity of the 1940s." *Working Paper*, WP-87-92. International Institute for Applied Systems Analysis: Laxenburg.

Keyfitz, N. (1988), "The Asian road to democracy." *Society*, 26(1):71–76.

Keyfitz, N. (1990), "The profile of intercohort increase." *Mathematical Population Studies*, 2(2):105–117.

Lutz, W., Goujon, A. and Doblhammer-Reiter, G. (1999), "Demographic dimensions in forecasting: adding education to age and sex." In: *Frontiers of Population Forecasting*, W. Lutz, J.W. Vaupel and D.A. Ahlburg, eds. pp 42–58. A Supplement to Vol. 24, 1998, Population and Development Review. Population Council: New York.

Lutz, W. and Goujon, A. (2001), "The world's changing human capital stock: multi-state population projections by educational attainment." *Population and Development Review*, 27(2):323–339.

Lutz, W., Sanderson, W. and Scherbov, S. (2001), "The end of world population growth." *Nature*, 412:543–545.

Lutz, W., Sanderson, W. and O'Neill, B. (2004), "Conceptualizing population in sustainable development: From 'Population Stabilization' to 'Population Balance." In: The end of world population growth in the 21st century: New challenges for human capital formation and sustainable development, W. Lutz, W. Sanderson, S. Scherbov, eds. pp. 315–334. Earthscan Publications, James and James: London.

Preston, S.H. and Coale, A.J. (1982), "Age structure, growth, attrition and accession: a new synthesis." *Population Index*, 48:217–259.

UN. (1999), *World Population Prospects. The 1998 Revision. Volume II: Sex and Age*, United Nations: New York.

World Commission on Environment and Development ("Brundtland Commission"). (1987), *Our Common Future*, Oxford University Press: Oxford.

CHAPTER 8. STRUCTURAL AND POLICY CONSEQUENCES OF MORTALITY AND FERTILITY DECLINE

SHRIPAD TULJAPURKAR
Stanford University, USA

Introduction

Demographic transitions have been a central issue in national and world policy for some time, and are well set to continue to play a key role in the future. At one time, the main transition of interest was fertility reduction in poor or developing countries. But at the start of the 21st century, many kinds of transitions have interest and significance: the transient and permanent effects of fertility decline, the transient impact of baby-booms or busts, long term declines in mortality, and fertility decline to levels well below replacement. Each of these has relevance in different places and times. To the extent that demographic change in currently developed nations is a prolog of what will come to pass elsewhere, the transitions just mentioned form a trajectory that will affect many nations over time.

The goal of this paper is to examine the early stages in the sequence by considering the policy consequences of changing population structure in the aftermath of fertility decline. The approach taken focuses on governmental expenditures on education, health and pensions. The focus on government is simply a reflection of the central role that governments play in the areas of human capital and welfare. This paper presents an analytically based method to combine economic and demographic information in order to understand and illuminate policy issues. The analysis does not directly address endogenous changes in demography and economy, although some of these can be taken into account indirectly, as will be indicated.

Our principal example is of a country that makes a fairly rapid transition, completed in less than about 15 years, from high fertility to low fertility. We focus on such rapid transitions because many Asian countries have in fact experienced transitions almost this fast, and because the analytical results are relatively straightforward in such a case. Slower transitions can be tracked by the methods discussed by Nan Li and Tuljapurkar (this volume). In our main example, the final state is one of replacement fertility – but this is not a necessary assumption for the analyses that follow. Just before the transition begins, the population is assumed to have a stable age distribution corresponding to the high fertility regime; again, this assumption may be relaxed. The economic and demographic data used are loosely based on India. But the analysis here is stylized and draws on new formal demographic methods (Li and Tuljapurkar 1999a, 1999b, this volume) – the machinery used to obtain our results is robust to changes in assumptions about the demographic details of a transition.

S. Tuljapurkar et al. (eds.), Population, Resources and Development, 139–154.

On the economic side, we discuss issues that have fairly general application, and the Indian example is illustrative rather than specific.

The method here has potential advantages over the use of traditional population projections, say those made by the UN or by national planning agencies. First, the present analysis rests on an understanding of the underlying formal dynamics, and so lends itself to exploring alternative assumptions about the trajectory of change of various factors over time. As discussed later the present approach has utility in assessing the uncertainty of projections. Second, the present analysis clearly indicates the importance of a long time horizon in the assessment of policy, by highlighting the way in which policy measures change in response to longer term demographic change. Third, the method is fairly easy to graft onto more detailed models that may be necessary to examine some aspects of population change. Fourth, in the context of this volume, the analytical method tracks changes in the age structure and numbers of a population over time so that the implications of age-structural transitions can be followed explicitly. The next section presents the demographic dynamics. The following two sections present the method and examine the economic and policy implications. Then we consider the effect of various supplements and extensions to the basic method outlined. The final section is a brief summary and discussion.

Demographic Dynamics

Consider a population that has been in a high fertility regime for some years and whose population has a stable distribution for that regime. Suppose that the population makes a transition to replacement fertility that is completed in 15 years. The formal dynamics of such a rapid transition have a great deal of simplicity – which is why we use it here. However, note that one could extend the analysis to study a 30 year transition or a 45 year transition without changing the essential details of what follows. For illustrative purposes, we use an initial age profile of fertility based on India's in 1971 and an initial Net Reproductive Rate (NRR) of 3, and a final age profile based on India's in 1992 and a final NRR of 1. The age profiles come from a time series of data from India's SRS, and though the shift between profiles is not huge it does show a substantial reduction in late-age childbearing. Fertility change is taken to be linear in time between the two age profiles, so that

$$f(a,t) = g(a) + h(a)\,k(t),$$

with $k(t)$ being a linear declining function of time. This type of parameterized model is easily adapted to describe many components of the processes we study. At present we take mortality to be fixed, and use a Coale-Demeny Model West life table (Coale and Demeny 1983) with an e0 of 65 years.

The age-structural shifts over a 75 year period from the start of the transition are best displayed graphically. The first 25 years (Figure 1) show that only the 0–5 age group sharply tracks the fertility decline – higher age groups take much longer to respond. Observe that the 15–20 age group starts to show a decline in numbers only at the end of the transition – this is a long tail that is being wagged. Over the first 50 years (Figure 2) most working

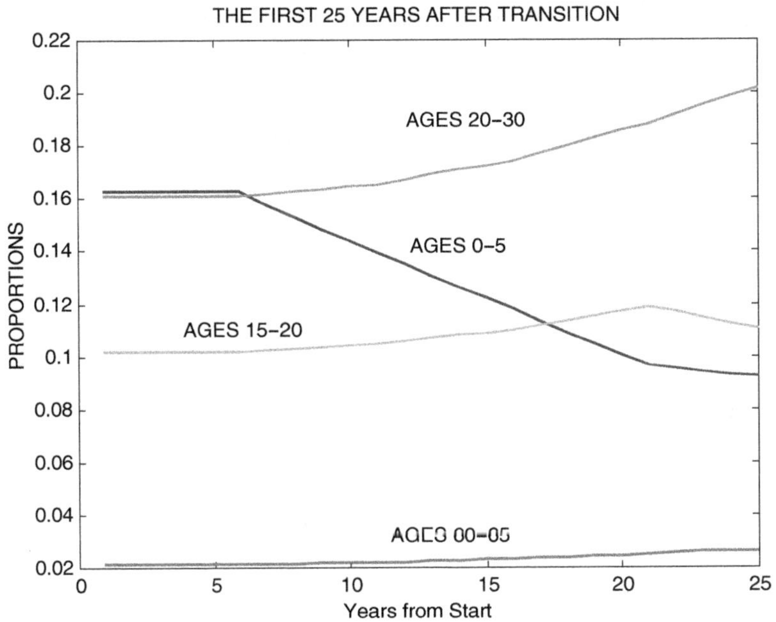

Figure 1.

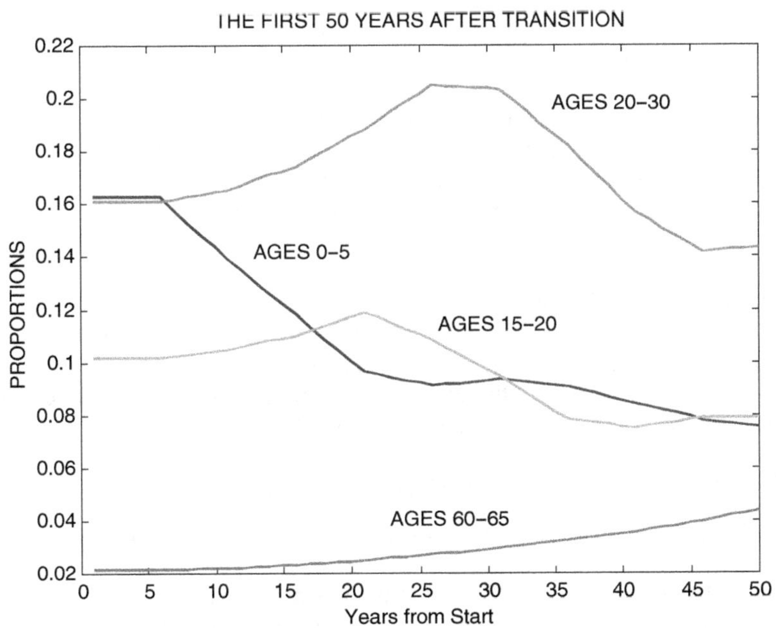

Figure 2.

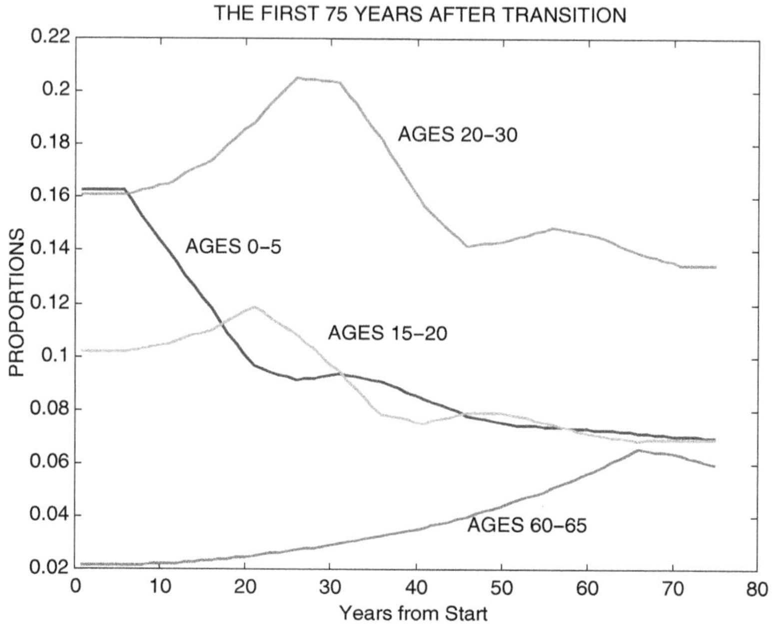

Figure 3.

ages reflect the transition. Over this interval, the 0–5 age group has declined sharply in number and shows a generational transient wave as it begins to stabilize towards its new stable level. The 15–20 and 20–30 age groups show similar patterns – there is a long rise to a peak followed by a decline towards a generation-long swing. The 60+ ages still continue their upward trend as large early cohorts march towards old age. Finally, 75 years after the transition begins (Figure 3) the 60–65 age group peaks and begins its decline. By this time, the cyclical change of the younger age groups has begun to damp out and stabilize.

In Figure 4, we show a still longer time period, but this time we present the three conventional age groups of life-cycle analysis: young (under 20), the potential labor pool (20 to 64) and the conventionally defined old (65+). Note the characteristic pattern in which the older age groups have a long period of growth before they go into an oscillatory decline. A different view of these numbers is provided by plotting dependency ratios:

YDR = Young Dependency Ratio = [Number under age 20]/[Number aged 20–64],

ODR = Old Dependency Ratio = [Number over age 65]/[Number aged 20–64],

and the Total Dependency Ratio (TDR) that is the sum of these. Figure 5 shows how the YDR declines rapidly to a minimum in about 60 years, and then shows a small oscillatory change. The ODR increases steadily before starting to stabilize after about 75 years. The TDR shows perhaps the most interesting behavior – initially it is dominated by the fall

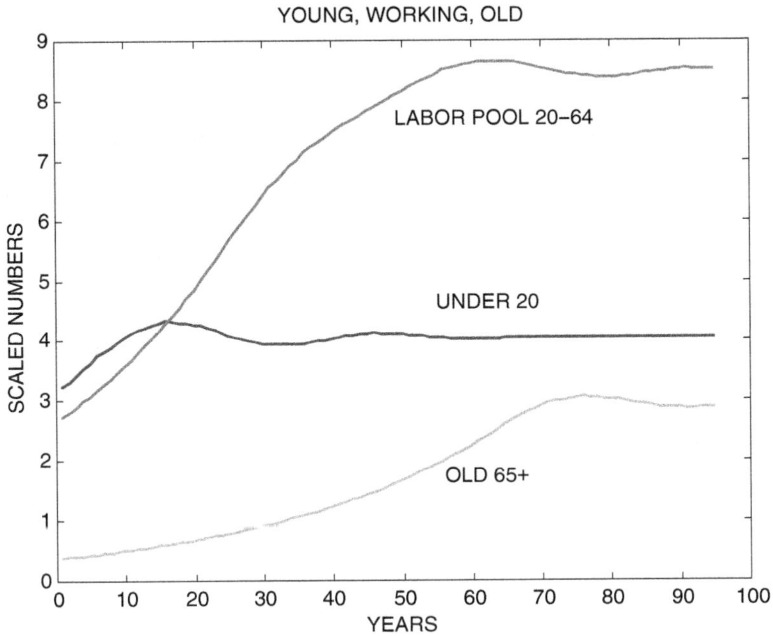

Figure 4.

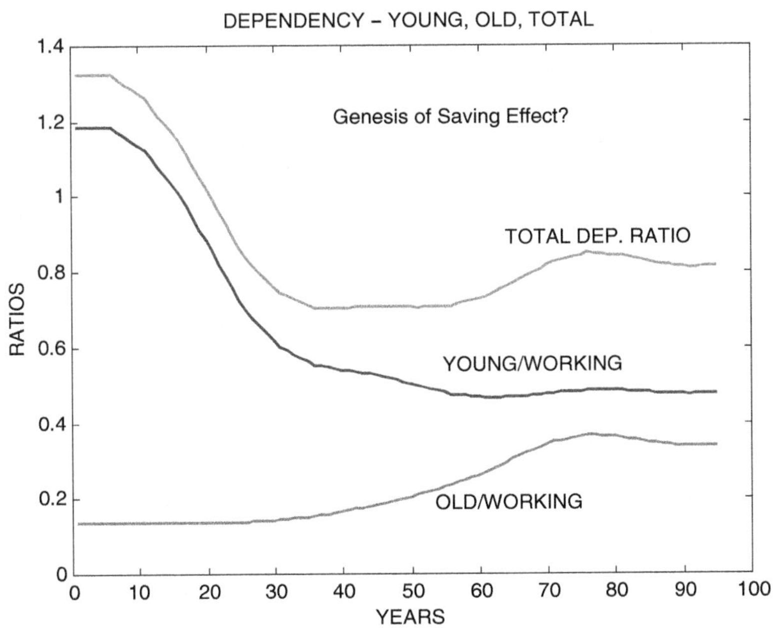

Figure 5.

in the YDR, but then it is noticeably pushed up by the increasing ODR before starting to stabilize in an oscillatory fashion.

The dependency ratios in Figure 5 provide a first, and important, point of contact with economic questions. Life-cycle analysis holds that the age profile of individual or household consumption and savings is based on a maximization of utility that takes into account one's consumption at all ages. Thus, if household heads have fewer children to support, they will respond by reducing consumption and increasing savings at earlier ages in their life cycle. At a population level, this argument suggests that savings rates track the inverse of the Young Dependency Ratio. Higgins and Williamson (1997), and Lee, Mason and Miller (1999, and elsewhere) argue persuasively that this phenomenon underlies the rapid growth of savings in many Southeast Asian countries in the past several decades.

The pattern of change we have illustrated here is characteristic. One may replace a linear fertility decline by a logistic trajectory, increase the duration of the transition from 15 to 30 years or longer, or incorporate slow declines in mortality – the broad patterns in the first 50 years after the transition will remain similar to those described here. This robustness is based on a formal analysis (Li and Tuljapurkar, this volume). The type of parametric model we use is easy to identify and provides a straightforward approach to constructing a suite of alternative projection scenarios.

Policy Consequences: Method

Here we consider policy as reflected in the trajectory of potential future government expenditures on education, health and pensions. These are a large part of what are called transfer payments to individuals, and are typically funded out of tax revenues. The method used is to estimate an age profile of per-capita expenditures in each category, call it $c(a,t)$ for age group a in year t for a particular type of expenditure. Total expenditures in that category depend on the population $n(a,t)$ at that age in that year, and are

$$C(a,t) = c(a,t)\,n(a,t).$$

The method combines demographic and other assumptions to examine how such transfers vary over time and age.

It will be useful to consider the sources of change in expenditure:

(1) Changes in $n(a,t)$ in the example of Section 2 result, of course, from the fertility transition. Other changes may well be important in particular cases, especially transitional or post-transitional shifts in mortality, fertility or immigration. We consider some of these later on.

(2) Some scale changes in the level of expenditure $c(a,t)$ are driven purely by a deepening effect of economic growth. As economies scale up in GDP, there will usually be a corresponding scaling up – a deepening – of expenditure levels, paid for out of increasing tax revenues. The main example here ignores these changes in level and concentrates on relative allocation patterns.

(3) Other scale changes in $c(a,t)$ are driven by changes in participation or coverage. This happens with educational expenditures when an increasing percentage of the population is provided with educational access or expanded opportunity, and in a similar way with health care and pensions. Again, the main example here ignores these effects, though we will later discuss how they may be included in the framework of this analysis.

(4) Some changes in the shape – the age-pattern – of $c(a,t)$ are driven by the nature of economic change. For example, educational spending at the university level or on technical training typically increases as economies become more industrialized. Such changes are discussed later in the paper.

(5) Finally, some changes in $n(a,t)$ or $c(a,t)$ are driven by the social and economic consequences of fertility decline. For example, lower fertility leads to decreasing family size and a thinning out of family and kin networks. These may require a greater governmental role in childcare, education, and elderly support. Changes in family structure can also increase the independence of young adults, affect marriage patterns, and set in motion longer term declines in fertility. We consider such issues briefly at the end of the paper.

For the present, then, focus on a situation where the per-capita expenditure profiles are held fixed, and changes in age-pattern of expenditure are driven by shifts in population structure.

Policy Consequences: An Example

The expenditure profiles used here are based on estimates made for India in 1981 by Lee (1991). Again, these are used because they are illustrative of what we expect in the early stages of development. Figure 6 illustrates the baseline – profiles of total transfer expenditure in three categories at the start of the transition. Note that education expenditures are concentrated on the 10 to 20 year age segment, and that both health and pension spending falls rapidly with age beyond age 60. The jagged nature of the profile for health spending reflects the use of a rough "step-function" that approximates spending on health by age.

Figure 7 shows that there are dramatic shifts in educational spending by age group. Spending on the young under 10 declines sharply with falling births, and then shows a cyclical upward echo followed by further decline. A similar but delayed pattern is seen at ages 10–20 and 20–30. These shifts imply substantial infrastructural adjustments over fairly short periods of time. The overall profile of educational spending by age, i.e., the relative total age-specific expenditures $C(a,t)$ measured relative to the youngest age group (Figure 8), shows the substantial shifting of expenditures by age over time. The changes seen here are in fact likely to be overshadowed by shifts in the pattern of demand for educational expenditures – a point to which I return later.

Figure 9 shows very substantial shifts in health-care spending, moving from spending dominated by young ages towards spending dominated by older ages. The sharp downward "tail" at high ages reflects our assumption of a fixed mortality schedule, and will be modified by an expected decline in mortality rates. Pension spending (Figure 10) also shows a strong

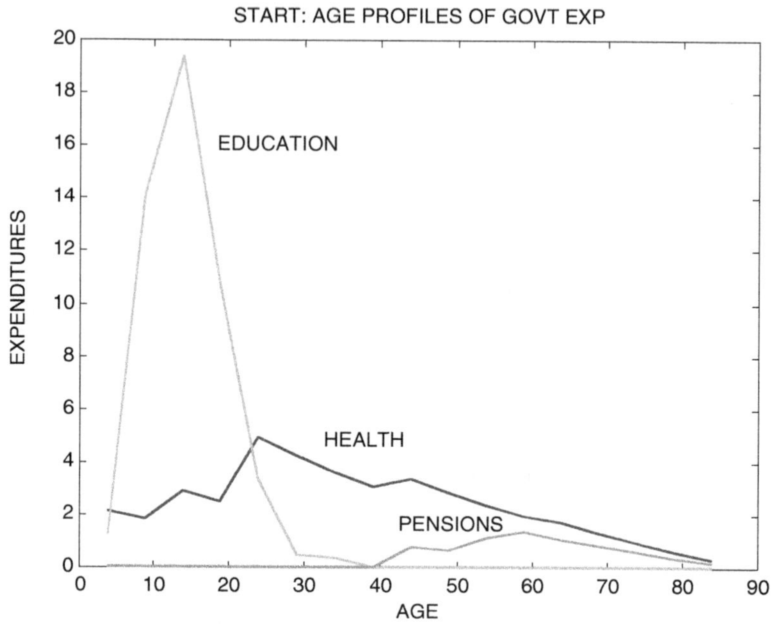

Figure 6.

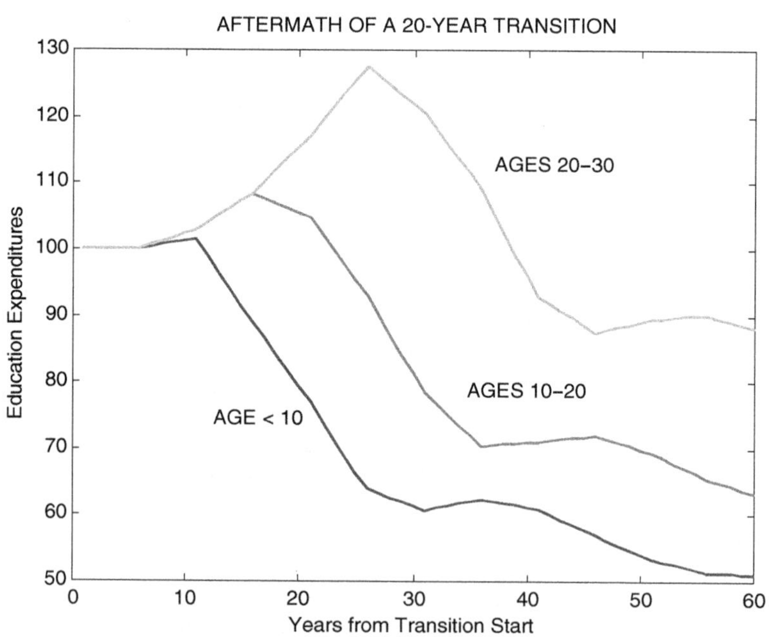

Figure 7.

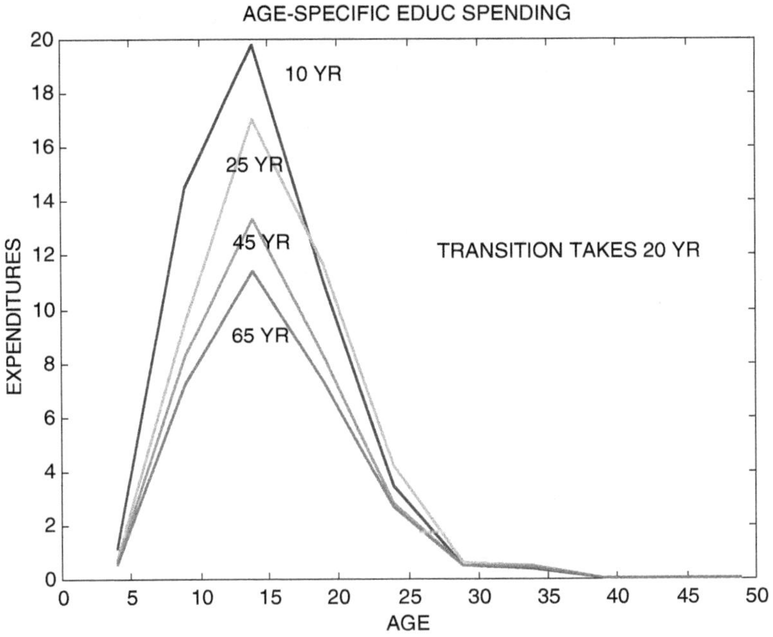

Figure 8.

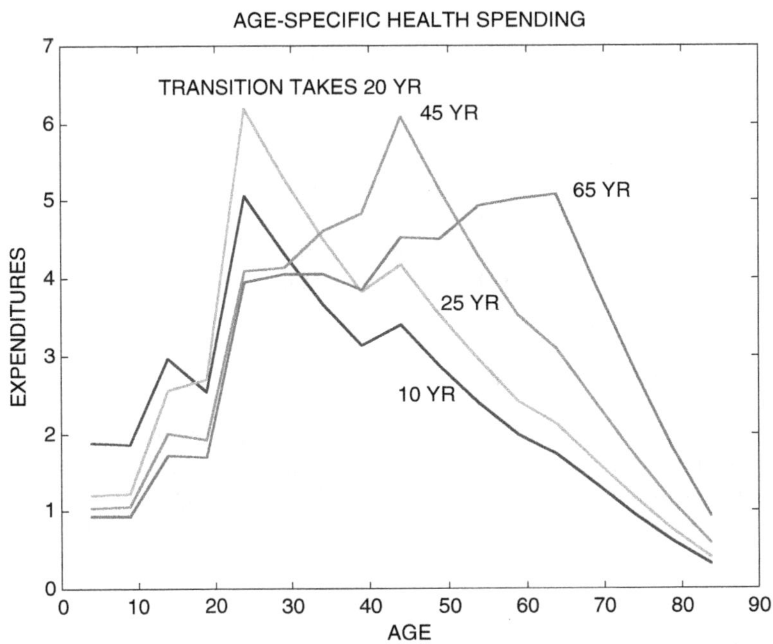

Figure 9.

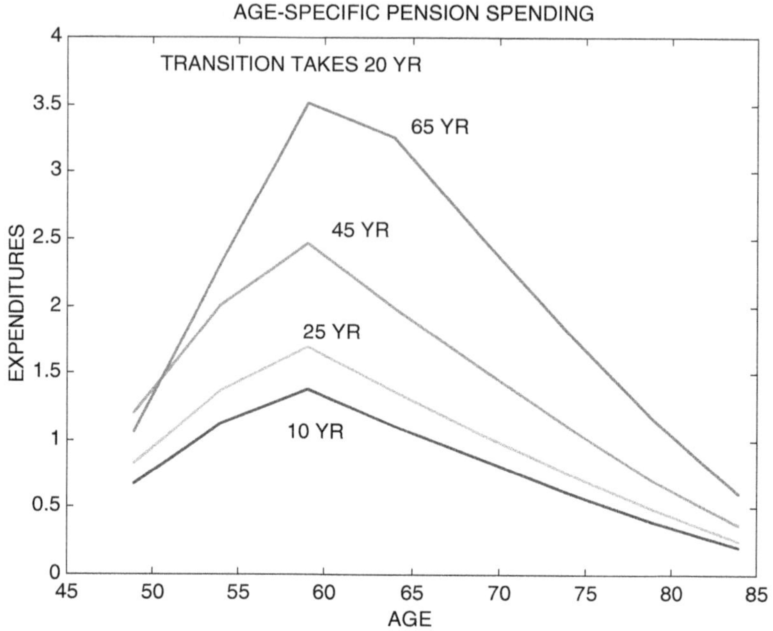

Figure 10.

shift to increased spending at older ages. Note that this shift occurs even in the absence of a large decline in mortality.

Similar shifts in all three age profiles continue as we move further in time after the transition. Note that over time the government must considerably increase resources to health and pensions: Figure 11 makes this point by plotting total expenditures on education, health and pensions over time. Note the steady upward march of the latter two categories, a harbinger of the longer term issues that currently occupy the world's richest countries.

Transitions Beyond Fertility Decline

We now turn to some important transitional features that are not included in the analysis thus far. The goal of this section will be to indicate what these transitions are and how they can be incorporated into the present framework. Even without explicit analysis, we can see how some of these will play out in terms of challenges to government policy.

EDUCATION

First, as pointed out earlier, the consumption patterns $c(a,t)$ are likely to evolve. Consider Figure 12, which shows the per-capita transfer profile for educational expenditures for India, 1981, and the US, 1994. To focus on the issue of allocation by age, rather than the

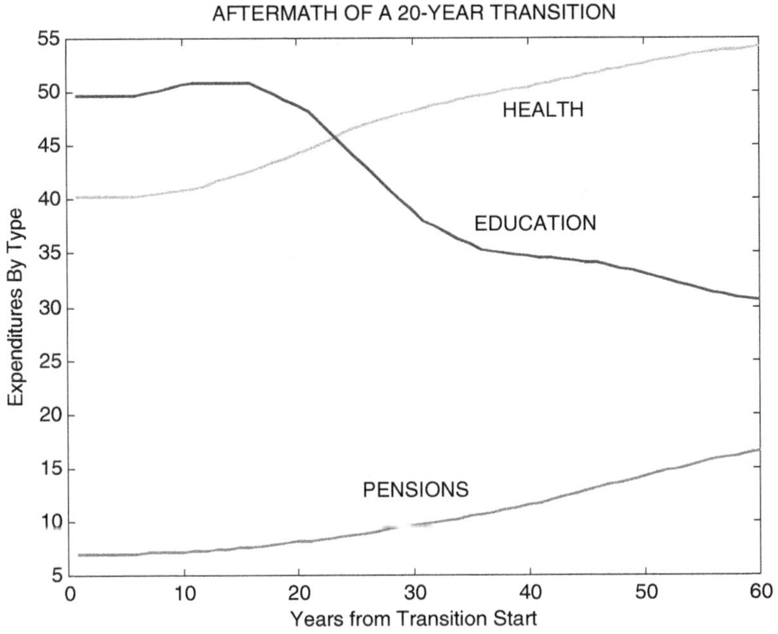

Figure 11.

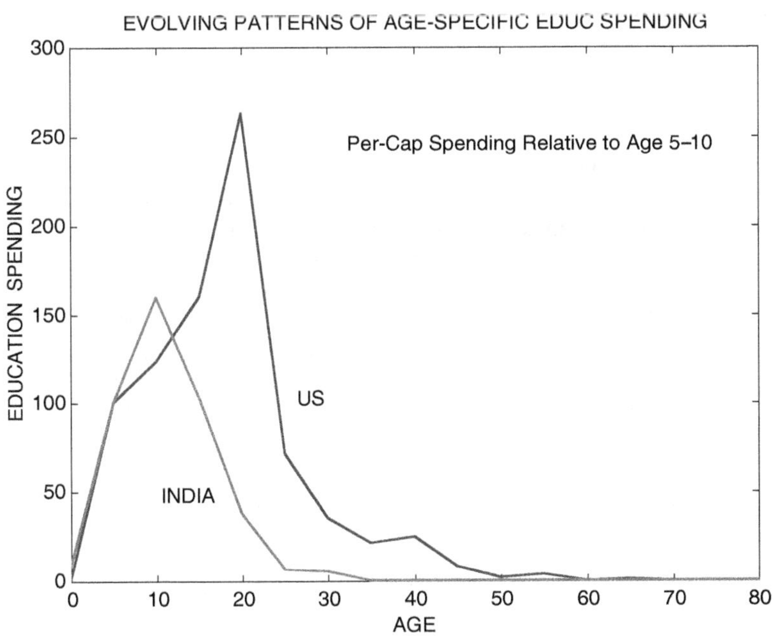

Figure 12.

substantial difference in absolute levels, the figure plots both profiles as curves relative to expenditure on the 5–10 age group. Observe that educational spending is substantially broader with respect to age in the US, with much greater emphasis on the older ages from 20 to 30. In fact, the US figures include state and federal expenses both on colleges and on support to individuals for educational expenditures. This broadening is what we expect as the level of industrialization increases. A second evolutionary aspect concerns the utilization of educational services. Along with development comes a strong push towards the wide, if not universal, availability of educational opportunity. A straightforward way of incorporating such shifts in the analysis would be to model the expenditure profiles as we did fertility profiles, by setting

$$c(a,t) = d(a) + r(a)\,s(t),$$

so that the time factor $s(t)$ scales between an initial and final pattern. For target patterns we could, as suggested here, use patterns from countries in the later stages of industrialization. Of course this procedure is complicated by the differing roles that the state plays in countries that have more versus less of a "socialist" policy perspective. But such differences are often less substantive than postural, and real differences can often be dealt with by properly accounting for transfers of various types. Looking back at Figures 7 and 8, then, we would now find that a broadening of the educational expenditure profile would strengthen and make permanent the shift in educational spending towards older ages. It would also increase the total commitment of transfer money. The relative time scales of a shift in expenditure via function $s(t)$ and in fertility via function $k(t)$ would determine the timing of these shifts and increases. These relative time scales will depend on the initial state of the country as regards industrialization and technology.

PENSIONS AND DECLINING MORTALITY

A second example concerns pension spending. Figure 13, top panel, shows per-capita pension expenditure profiles for India 1981 and the US 1994. Again, we plot each of these relative to spending at age 65. Notice that in India retirement ages are fairly low, although they have shown an upward trend in more recent years. One reason for this age pattern is that most Indian pensions are for people leaving government service, in which many pensions were (and often are) payable upon completion of a specified term of service (such as 20 years) and not necessarily upon reaching a set age of retirement. The US profile mainly considers payment out of the now nearly universal Social Security program, and so is an underestimate of total transfers to retirees. Trends in India and elsewhere suggest that the retirement pattern is evolving in the direction of that in the US, i.e., to later ages of retirement. In addition, after a transition to low fertility, populations display a strong shift towards weaker family support of the elderly – this puts pressure on governments to step into the breach. Typically, there is an expansion of public support programs for the elderly, along with initiatives to expand privately funded pensions.

These changes occur against a background of secular decline in mortality rates. Figure 13, bottom panel, shows the central death rates (log scale) for India and the US both in 1992. Analysis of time trends (not presented here) strongly suggests that Indian mortality rates are moving downwards at a slow but persistent pace. By analogy with methods we have

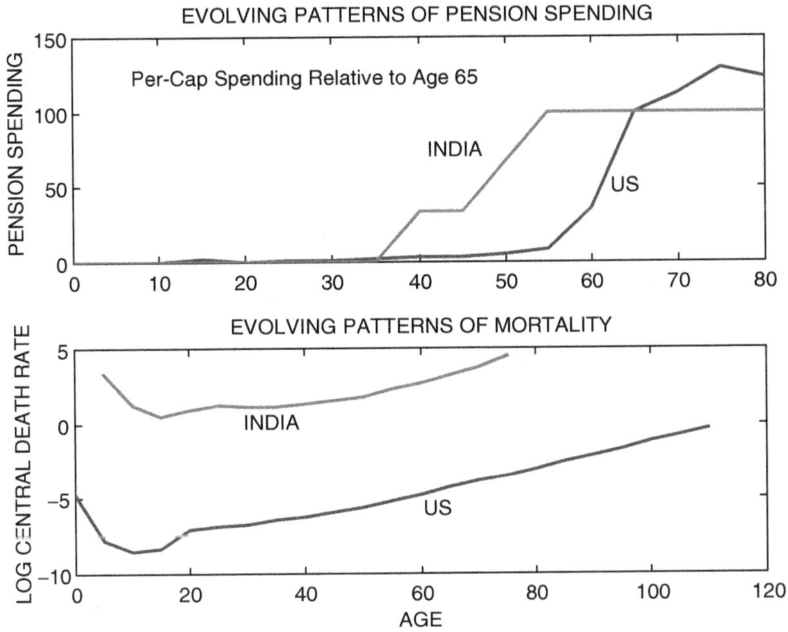

Figure 13.

used elsewhere for mortality forecasting (Tuljapurkar, Li and Boe 2000), one could model changes in central death rates $m(a,t)$ by

$$\log m(a,t) = \log m(a,0) + B(a)\,K(t).$$

In this case we would need to choose a "target" set of mortality rates – alternatively, one could choose a $B(a)$ and $K(t)$ by comparative analysis. The evolution of the pension spending profile could be modeled similarly.

Clearly, as a result of the combined effect of changes in the pension profile and reduced mortality such evolution will, in most cases, make future pensions appear to be a major potential challenge for fiscal policy. There are several virtues to such a modeling exercise, however. First, it provides early warning of the prospects. Second, it suggests how government could respond by attempting to shift retirement behavior, savings patterns, and private versus public pension programs. Third, it can usefully inform policy options regarding public financing of pensions, vis-à-vis the role of individual and corporate contributions to pension plans.

Health-care spending also has evolutionary aspects but these are more difficult to model because of large systematic differences in health-care delivery and consumption. However, the role of government as a key payer for many aspects of health care is widely established, and indeed essential. Here too, we would look for a large amplification of the trend to higher expenditures that was displayed earlier.

Discussion

EXTENSIONS TO OTHER FEATURES OF TRANSITION

There are several issues that have not been considered here but are important. One is population heterogeneity, and its changing role in the course of transitions. Fertility transitions are well known to be heterogeneous in space and time within nations, and thus the changes that we describe are likely to affect different parts of a country at different times. This can be a major problem in terms of resource allocation, and deserves attention. Another source of spatial heterogeneity is the urban–rural gap that is typically widened by development, and is often amplified by rural migration into urban areas. Indeed the latter shift continues at all stages of development and poses a range of challenges along the entire trajectory of transitions. Spatial variation can be addressed by using spatially explicit, multiregional, versions of the present models (Rogers 1989). There is also heterogeneity in terms of socioeconomic inequality and access to opportunity, within and between spatial regions. The theoretical analysis of such inequalities has begun to address the forces that maintain inequality, as discussed for example by Sen (2000). In this context, demographic analysis can be a useful complement to other types of analysis, mainly by showing how inequalities can expand or contract as a result of demographic shifts.

Migration, both into urban areas, and international, can be an important issue for some countries. In the lowest fertility countries, large net immigration is increasingly seen as one of the solutions to rapid population ageing. In higher fertility countries, temporary and permanent migrants to other countries can be a significant source of remittance earnings, as well as a siphon of skilled labor. In some countries, such as China in recent years, large and rapid migrations in response to employment opportunity have become a distinctive feature, and pose special problems in terms of the provision of services and regional development. Migration can be addressed using multiregional models. We have mentioned some of the social impacts of transitions on family, kinship, marriage, and traditional support systems. The modeling of such aspects requires far more detailed models than we have used, such as the SOCSIM model at Berkeley (Wachter 2004). However such models can be used to supplement and complement the present analysis so as to explore the more detailed demographic structures that are involved. A related social aspect of transitions is the impact of low fertility and mortality regimes on the male–female sex ratio at early ages and the "masculinization" of the sex ratio. Tuljapurkar, Li and Feldman (1997) showed how one can model the impact of such sex ratio changes on the short run dynamics of marriage, but further analysis of the dynamics of this question are desirable. It seems clear that mortality and fertility decline can cause substantial distortions of social arrangements that are thought of, traditionally, as equilibrium arrangements, and it would be useful to examine the new equilibria that are likely to supplant the old.

UNCERTAINTY

There are many arguments for the development and application of an explicit probabilistic analysis of uncertainty in demographic projections (Tuljapurkar and Lee 1998). Traditional scenario alternatives are useful but are limited in an essential and important way – they do not provide a useful or consistent description of the relative likelihood of various outcomes. We are not faulting here the methods by which scenarios are constructed – even though, like all methods, they can usually benefit by scrutiny and change. Rather, it is important to notice that the rich underlying analyses, objective and subjective, that go into developing scenarios are obscured by the conventional presentation of a dominant scenario and a few unlikely alternatives. It is considerably more useful to provide a clear indication of the relative probabilities that attach to any possible outcome. This can be done by using explicitly probabilistic methods at every stage of the development of models. The resulting models are quite robust to tinkering with one or a few factors. They allow analysts to compare different objective functions in a consistent way, and to assess the risks that accompany particular policy choices.

Conclusion

As one looks at evolutionary patterns in education, health and pensions, the issues that stand out are in fact precisely those that occupy the highly industrialized countries today. As suggested at the beginning of this paper, the course of transitional shifts seen in different places suggests some degree of natural order which can be exploited in analyses of the sort suggested here. Many highly industrialized countries are now also affected by the second, and perhaps last, fertility transition – to fertility levels well below replacement. To the extent that this transition persists, our analysis may be applied with the assumption that the final state is one of steady decline in births accompanied by a steady expansion of lifespan.

The approach exemplified here allows us to put a substantive face on the implications of age-structural shifts for policy, as discussed by Ian Pool (this volume). It shows how relatively short term (10 to 30 year) population shifts can cause substantial challenges for governments in terms of the relative demand for expenditures on critical elements of human development and support. It also provides a strong sense of the way in which longer term (30 to 75 year) trends are likely to challenge societies in the future.

Acknowledgements

This work was supported by Mountain View Research. I thank Ian Pool and Anne Gauthier for stimulating discussions that led to this analysis, and Nan Li, who has worked closely with me on many aspects of population transitions.

References

Coale, A. and Demeny, P. (1983), *Regional Model Life Tables and Stable Population*, Academic Press: New York.

Deaton, A. and Paxson, C. (1998), “Saving, growth and aging in Taiwan.” In: *Studies in the Economics of Aging*, D.A. Wise, ed. University of Chicago Press: Chicago, IL.

Higgins, M. and Williamson, J.G. (1997), “Age structure dynamics in Asia and dependence on foreign capital.” *Population and Development Review*, 2:261–293.

Lee, R.D. (1994), "The formal demography of population aging, transfers, and the economic life cycle." In: *The Demography of Aging*, L. Martin and S. Preston, eds. National Academy Press: Washington, DC.

Lee, R.D., Mason, A. and Miller, T. (1999), "Saving wealth and population." Presented at the *1999 Annual Meeting*, Population Association of America: New York.

Li, N. and Tuljapurkar, S. (1999), "*Population momentum.*" *Population Studies*, 53:255–262.

Li, N. and Tuljapurkar, S. (1999), "The Solution of time-dependent population models." *Mathematical Population Studies*, 7:311–329.

Li, N. and Tuljapurkar, S. "A formal model of age structure transitions." (this volume)

Pool, I. "Age-structural transitions and policy: frameworks." (this volume). Rogers, A. 1989, *Multiregional Mathematical Demography*.

Sen, A. (2000), *Development as Freedom*, Alfred A Knopf: New York.

Tuljapurkar, S. and Lee, R. (1998). "Demographic uncertainty, risk analysis and public policy." *Proceedings of the Bernoulli Society*.

Tuljapurkar, S., Li, N. and Boe, C. (2000), "A universal pattern of mortality decline in the G7 countries." *Nature*, 405:789–792.

Wachter, K. (2004), "The Berkeley SOCSIM model." *Working Paper and Manual*, University of California: Berkeley. Online at http://www.demog.berkeley.edu/~wachter

Part III

COUNTRY-SPECIFIC TRANSITIONS AND CHALLENGES

CHAPTER 9. POLICY IMPLICATIONS FOR OLD-AGE ECONOMIC SUPPORT OF CHANGES IN THAILAND'S AGE STRUCTURE: A NEW CHALLENGE

NAPAPORN CHAYOVAN
Chulalongkorn University Thailand

Introduction

In the new millennium, the demographic shift from a younger to an older population age structure will be a universal feature of the world's populations, including Thailand. In most industrialized countries this process has occurred gradually, taking up to a century, which has allowed societies and economies to adapt to their demographic evolution. In Thailand, this process has been rapid and will continue to accelerate. The rapidity of the current change in Thailand stands in contrast to historical developments in Western countries. This implies that Thailand will face emergent issues related to social security, health care costs and intergenerational equity in a far shorter time span than was true in the West. The development of appropriate policies and programs to tackle these issues is required.

This paper attempts to demonstrate the future burden, for the family and society, of an older population and their needs. The paper begins with an overview of the population ageing process, followed by a description of the elderly population's economic circumstances. Trends in labour-force participation, income level, debt, savings, income sufficiency and satisfaction, and the main sources of income are presented. Preparation for old age among the elderly and the near elderly will be discussed in order to highlight the need for policies and programs which will ensure the economic well-being of the future elderly. In this paper, elderly is defined as those who are 60 years old and over.

Data Sources

Most of the data provided in the paper are from the two major national surveys of the elderly: the 1986 Socio-Economic Consequences of the Ageing of the Population in Thailand (SECAPT); and, the 1995 Survey of the Welfare of the Elderly in Thailand (SWET). Both surveys were conducted by the College of Population Studies of Chulalongkorn University, formally known as the Institute of Population Studies. The SECAPT study interviewed 3,252 respondents and SWET interviewed 4,486 respondents aged 60 and over nationwide. In addition, 3,222 persons aged 50–59 were also interviewed in SWET. Details of the research design and methodology of the SECAPT and SWET are described

S. Tuljapurkar et al. (eds.), Population, Resources and Development, 157–179.

in the survey reports of the two projects (Chayovan et al. 1988; Chayovan and Knodel 1997).

Demographic Change and the Pace of Population Ageing

During the past several decades, Thailand has proven to be one of the most successful countries in bringing down its fertility level within a short period of time. The total fertility rate has declined from over 6 births per woman in the mid 1960s to below 2 in the mid 1990s (NSO 1997). Life expectancy at birth increased from 55.2 years to 69.9 years for men and 61.8 years to 74.9 years for women during the same period. Besides lowering the growth rate, a major demographic consequence of this rapid fertility reduction will be an inevitable ageing of the population in the coming decades. Even more dramatic will be the rapid increase in the absolute size of the elderly population (aged 60 and over), a result of past high fertility levels and substantial declines in mortality.

Table 1 presents the past trends and future projections of key demographic indicators related to population ageing in Thailand for the period 1980–2050 as estimated by the United Nations (United Nations 1999). Three main characteristics of the ageing population in Thailand are: a considerable change in the age structure of the population with a rapid increase in the share of the elderly population; the ageing of the elderly population itself; and, an increase in the proportion of females in the older age groups.

Trends in the growth rates and sheer size of the elderly population are also remarkable. The elderly Thai population is growing faster than the growth of the total population. The rapid increase in the proportion of the elderly population implies a sharp increase in the size of the elderly population. Thailand's total population grew by 31 percent between 1980 and 2000, increasing from 46.7 million to just over 60 million at present. At the same time the population growth declined substantially, from 3 percent around 1970 to under 1 percent currently.

The proportion of the population in their elderly years (60+) is anticipated to increase from 8.7 percent in the year 2000 to 10.8 percent in the year 2010, 15.2 percent in the year 2020, and 30 percent in the year 2050. The number of older persons will continue to increase, from approximately 5.3 million at present to 7.2 million in 2010 and will reach 11 million by 2020. Based on the latest projections from the United Nations, the growth rate of the Thai elderly population is relatively high, over 3 percent per year. With the growth rate at 3–3.6 percent per year, the size of the elderly population will double every 19–23 years or so. Thailand will become an ageing society within the next 10 years, according to the United Nations' definition.

Not only is the elderly's overall share of the population increasing, but the elderly population itself is also ageing as evidenced by a rise in the percentage of the elderly who are aged 70+ or 75+. The percentage increase of the old-olds is greater than that of the overall aged population. The sex composition of the elderly population appears to be relatively stable from 1990–2020. There are more elderly women than elderly men in every age

Table 1. Projected trends in selected demographic measures of the elderly population in Thailand, 1980–2050.

	1980	1990	2000	2010	2020	2050
1) Population (in 1,000)						
Total	46,718	55,595	61,399	66,511	70,975	74,188
60+	2,527	3,716	5,338	7,205	10,765	21,981
65+	1,649	2,408	3,576	4,953	7,168	17,076
75+	484	798	1,192	1,852	2,594	8,222
85+	–	–	198	327	540	1,946
2) % Increase over 1980						
Total	–	19.0	31.4	42.4	51.9	58.8
60+	–	47.1	111.2	185.1	326.0	769.8
65+	–	46.0	116.9	200.4	334.7	935.5
75+	–	64.9	146.3	282.6	436.0	1,598.8
3) Population growth rate						
Total		1.7	1.0	0.8	0.6	0.1*
60+		3.9	3.6	3.0	4.0	0.9*
4) % of total population aged:						
<15	40.0	31.9	25.2	21.6	19.8	16.8
60+	5.4	6.7	8.7	10.8	15.2	29.6
65+	3.5	4.3	5.8	7.4	10.1	23.0
5) % of the total elderly population aged:						
70+	38.3	39.6	41.3	45.1	41.3	57.3
75+	19.2	21.5	22.3	25.7	24.1	37.4
85+	–	–	3.7	4.5	5.0	8.9
6) % Female						
60+	54.7	54.8	55.5	55.8	55.6	54.9
75+	58.5	59.9	60.0	60.7	60.6	59.0
85+	–	–	65.2	65.4	65.4	63.7
7) Dependency ratio						
Total	83.2	62.9	51.3	47.9	53.8	86.5
<15	73.3	52.0	38.1	31.9	30.5	33.2
60+	9.9	10.9	13.2	16.0	23.3	55.3

Source: Calculated from data provided in the United Nations (1999), *The Age and Sex Distribution of the World Population, The 1998 Revision*, pp. 784–785.

* Growth rate is an average of the years 2040–2050.

group. The proportion of elderly females is greater at older ages (panel 6, Table 1). This is because women are more likely than men to survive to older ages. The higher proportion of females at older ages implies a greater increase in the number of elderly women than elderly men. Any of the very old would be women, often widowed and without adequate means of support. As of 1995, widowed females constituted about one-fourth of the elderly in Thailand. Older women are likely to live alone, have poor health and face financial difficulties (Chayovan and Knodel 1997). Thus they are considered more vulnerable and deserve special attention and assistance.

Trends in the Dependency Ratio

The dependency ratio is a numerical measure of the economic burden imposed on the working population who must ultimately support people who are not in the labour force. The old age dependency ratio is an indicator that roughly quantifies the demographic weight of burden that the current working-age population has to bear in order to support the elderly. Changes in age structure that are taking place in Thailand translate into changes in dependency ratios defined in terms of age groups. In the past, children constituted the major part of the dependency ratio, but this is changing and by 2050 about 64 percent of the dependents will be the older adults. The increasing dependency burden of ageing populations would be more than offset by the falling dependency burden of young people during the early stages of ageing (1990–2000). The total dependency ratio has declined and will reach the lowest level in 2010, but will rise again in the second decade of the next century. The aged dependency levels, however, will steadily increase from about 13 elderly per 100 population aged 15–59 at present to 16 per 100 in 2010 and 23 per 100 in 2020 (panel 7, Table 1). The increase in the old-age dependency ratio implies that the burden of support for older persons will become heavier. This includes health care costs and other social and economic supports.

There can be no doubt that the altered age structure that is emerging as a consequence of fertility decline, and particularly the associated population ageing and future rapid growth of the elderly in Thailand, has important demographic and socioeconomic implications for the intermediate and long term future well-being of society. In particular, concern has been expressed about the potential erosion of existing family support systems, the increase in demand for health care, and the persistence of poverty among broad segments of the population. Not all of these demographic changes need be seen as negative, however, especially if appropriate planning and policies are implemented in a timely fashion to help in the adjustment process.

Speed of Population Ageing

The rapidity of population ageing in Thailand and some newly completed demographic transition countries is alarming. The number of years it took to shift the proportion of the elderly population from 7 percent to 14 percent is much lower in Thailand than it was in many industrialized countries. As shown in Table 2, it took France 114 years, Sweden 85 years and Italy 63 years to shift from having 7 percent of its population in the 65 and over age group to having 14 percent in that category. In comparison, it took Japan only 26 years to make that change. But now Japan has serious competitors in Asia with Thailand and Singapore both expected to take fewer than 25 years to make the transition. The shorter time Thailand will take to become an ageing society means that the country also has a shorter time to adjust to and make plans for this rapid demographic change. Population ageing is widely considered a cause of various social and economic problems. The increase

Table 2. Speed of population ageing in selected countries.

	Year the proportion of population aged 65+ is 7%	Year the proportion of population aged 65+ is 14%	Number of years required for the proportion of elderly to increase from 7% to 14%
Developed countries			
France	1865[1]	1979[2]	114
Sweden	1886[1]	1971[2]	85
United States	1941[1]	2013[2]	72
Italy	1924[1]	1987[2]	63
Japan	1969[2]	1994[2]	26
Asian countries			
South Korea	2001[2]	2023[2]	22
Singapore	2000[2]	2017[2]	17
Thailand	2007[2]	2029[2]	22
China (excl. Hong Kong)	2002[2]	2027[2]	25

Source: 1. United Nations, *The Ageing of Populations and Its Economic and Social Implication*, 1956, p. 12. 2. United Nations, *World Population Prospects: The 1998 Revision*, volume 1: Comprehensive table, 1999.

in the aged-dependency ratios will lead to problems of economic support and adequate health service provision for older persons.

Labour Force Participation: Level and Trends

The job and work patterns of individuals have lifetime implications. The impact of population ageing on society and the economy is, in part, a result of changes in the age structure of the labour force. As the population ages, the proportion of older persons who are economically active tends to decline. Labour-force participation of the elderly can thus be used as an indicator to reflect the extent to which the elderly are economically independent and the extent of their economic dependence on society. Time trends in the labour-force participation of older persons in Thailand are shown in Table 3 which presents the proportion working last week of the Thai elderly by age group, sex and residence for 1986 and 1995. Results indicate a substantial increase in the economically active rates during 1986 and 1995. Approximately 30 percent of elderly in Thailand worked in the last week before the survey in 1986 and the level had increased to 37 percent in 1995, representing a 25 percent increase during the 9 year period. In contrast, the experience of industrialized developed countries shows a trend towards a decrease in the work participation of older populations, but the decline was less for females than males (Kinsella and Gist 1995). The increase in labour-force participation of Thai elderly could be explained partly by Thailand's economic systems and situation during that period. Since Thailand is still largely an agricultural society and retirement is an alien concept, people tend to work until they are physically unable to do so, because of their financial needs. It is thus not surprising

Table 3. Labour-force participation rates (last week) of the elderly by age groups, sex and residence, 1986 and 1995, Thailand.

	Total		Residence		Urban		Rural	
Age group/year	**Male**	**Female**	**Urban**	**Rural**	**Male**	**Female**	**Male**	**Female**
1986[a]								
Age group								
60–64	54.6	34.5	27.2	45.9	39.6	20.0	57.7	37.8
65–69	43.5	28.6	28.6	36.8	31.5	26.4	46.1	29.1
70–74	29.2	15.3	14.5	21.9	23.1	9.6	30.3	16.4
75–79	25.0	9.5	12.8	16.7	19.4	8.8	26.0	9.6
80–84	18.1	9.0	9.4	13.3	15.6	5.5	18.6	9.9
85+	11.6	0.0	(0.0)	3.8	(0.0)	0.0	13.6	0.0
Total	**40.1**	**23.4**	**21.8**	**32.0**	**29.7**	**16.9**	**42.3**	**24.9**
1995[b]								
Age group								
60–64	66.2	47.9	42.2	59.9	51.0	35.5	68.9	50.8
65–69	56.1	30.0	30.6	43.8	36.2	26.6	60.1	30.7
70–74	33.1	18.5	18.8	26.8	18.0	19.0	36.5	18.4
75–79	18.9	14.2	8.3	18.0	3.5	11.2	22.3	14.9
80–84	14.2	4.6	6.0	9.2	12.4	1.7	14.7	5.4
85+	8.4	0.4	1.5	3.8	(0.0)	2.0	9.6	0.0
Total	**48.5**	**28.9**	**26.8**	**40.3**	**31.8**	**23.3**	**51.8**	**30.3**

Source: a) 1986 SECAPT; b) 1995 SWET.

to find that about 4 out of 10 elderly were still currently working in 1995. In addition, during 1986–1995 Thailand was experiencing an economic boom, unemployment was low and it was relatively easy to find a job. Results from labour-force surveys in Thailand, however, reveal a slight decline in the labour-force participation rates, from 52 percent to 48 percent for elderly males and 26 percent to 25 percent for elderly females (NSO 1988; NSO 1996). The difference in results may be due to variations in the definition used, or how the question was asked. Given the economic downturn the country has been facing since 1997, it is interesting to follow the future trends in the labour-force participation of the older population.

As expected, the labour-force participation rates of the older population decline as their ages increase. This is true regardless of sex and residence. For every age group, the rural elderly display higher rates of labour-force participation than do their urban counterparts. This may be attributed to the nature of work they had engaged in before reaching old age. Agriculture was the longest held occupation for rural elderly; while for urban elderly, commerce, unskilled and skilled labour, professional and clerical work are more dominant (Chayovan and Knodel 1997). It is common for people who work in agriculture to continue to work until they are physically unable to. In contrast, retirement age is usually imposed on workers in the formal sector but not for those in family owned commerce.

Gender differentials in labour-force participation rates are also evident. It is generally observed that female labour-force participation rates are lower than those of males for every age group. This is because women are traditionally expected to be housewives or full-time homemakers. In Thailand, although the overall labour-force participation rates for Thai women in the working age groups are relatively high compared to rates for women in many other countries, the females' rates are also lower than that of males. Thai women, however, are known to play an important role in managing household finances; their contributions can thus be more than working outside the house. Although the labour-force participation level of women is lower, the rise in the rates appears to be greater for women than for men.

In sum, although some elderly are still working, a majority of them are not. The main reasons cited for not working by more than 80 percent of the non-working elderly include old age and health problems (Chayovan and Knodel 1997). This implies that older persons need sources of financial support other than from their own work.

Desire to Work and the Unmet Need for Work

A strategy often proposed to improve the quality of life of the elderly is to facilitate their continuing involvement in productive activities extending their working life. However, the feasibility of such a strategy is difficult to judge without empirical data to assess it. Desire to work is subjective and should be distinguished from the objective potential to work. In designing a program to promote productive ageing, consideration should be given to the extent the elderly are able and willing to work. The main reason leading the elderly to stop working is their deteriorating health (Chayovan et al. 1988; Chayovan and Knodel 1997). Elderly people who are unwell would not be able to work even if they wanted to. Health status is thus likely to be associated with the desire and the ability to work. Elderly with good health are more likely to want and be able to work than those with poor health.

The associations between desire to work, ability to work and current work status (as measured either last week or last year); partly reflect an unmet need for work. The group with an 'unmet need' for work is defined in this paper as those who want to and are physically able to work but are not currently working (either last week or last year).

In SECAPT, respondents were asked if they wanted to work but were not directly asked if they were physically able to work; whereas SWET specifically asked the respondents if they wanted to work and whether they were physically able to work. The physical and mental capability to work may be determined by self-assessed health status. If respondents reported poor health they were classified as unable to work. Using this classification, the 'unmet need' for work was constructed for both 1986 and 1995. However, since SWET has direct information on the desire and the ability to work, an index of 'unmet need' was also constructed using this information. Results of the 'unmet need' index using direct information on the ability to work show a much lower level of unmet need than that using indirect information.

Table 4 presents the percentage of elderly who wanted to work and the unmet need for work identified last week and last year, by sex and residence for 1986 and 1995. Results

Table 4. Percentage wanting to work, and unmet need for work, during last week and last year among elderly by sex and residence, 1986 and 1995, Thailand.

	Total	Total		Residence		Urban		Rural	
Age group/year	**Total**	**Male**	**Female**	**Urban**	**Rural**	**Male**	**Female**	**Male**	**Female**
1986[a]									
% Wanting to work	65.6	71.6	61.5	51.0	68.6	55.6	48.1	74.7	64.4
% unmet need for work last week[1]	27.5	25.8	28.6	21.0	28.8	17.7	23.1	27.4	29.8
% unmet need for work last year[1]	23.5	20.6	25.4	19.8	24.2	17.0	21.5	21.3	26.3
1995[b]									
% Wanting to work	67.2	72.2	63.0	53.1	70.2	57.7	49.8	74.9	66.1
% unmet need for work last week[1]	9.2	8.8	9.7	13.9	8.3	16.3	12.2	7.3	9.1
% unmet need for work last week[2]	4.4	5.4	3.6	5.4	4.2	6.7	4.5	5.2	3.4
% unmet need for work last year[1]	7.0	6.1	7.8	12.5	5.8	14.6	11.0	4.4	7.0
% unmet need for work last year[2]	1.9	2.4	1.4	3.9	1.4	5.3	2.9	1.9	1.0

Source: a) 1986 SECAPT; b) 1995 SWET.

Notes: 1. Unmet need for work was constructed based on the information on current health status and desire to work and current (last week or last year) employment status.

2. Unmet need for work was constructed based on the direct information provided by the elderly respondents on their desire and ability to work and current (either last week or last year) employment status.

show that the desire to work for 1986 and 1995 are almost the same. About 67 percent of the elderly want to work. Elderly men are more likely to want to work than elderly women. This may be due in part to expected gender roles. Men are expected to be the breadwinners and thus have to work.

For both men and women, the proportion wanting to work is higher for rural elderly than urban elderly. As previously stated, Thailand is still an agricultural society and rural people are relatively poor. They may want to continue working into old age because of financial needs.

Results in Table 4 suggest a decline in the levels of unmet need for work. The proportions of elderly who wanted and were able to work but were not working in 1986 are considerably higher than in 1995. Results of the unmet-need measure using indirect information on the ability to work reveal that approximately 28 percent and 9 percent of the elderly in 1986 and 1995, respectively, wanted and were able to work but were not working last week. This substantial decline in the unmet need for work could be explained largely by the increase in the work opportunities during the period of economic boom.

In 1986, the level of unmet need for work among men was lower than for women regardless of residence. Among both men and women, the rural elderly demonstrate a higher level of unmet need for work than the urban elderly, reflecting the need to work for income among the former group. Out-migration of the young may contribute to the increase in job availability and the need to work among rural elderly. Using the comparable measure of unmet need for work in 1995, gender and urban–rural differentials in the level of unmet need for work are found to be opposite to the level in 1986. In 1995, gender differences in the level of unmet need for work are minimal and vary by residence. Overall, the level of unmet need for work was higher in urban areas than in rural areas. In urban areas, elderly males display a higher level of unmet need for work than elderly females, but in rural areas elderly females still show a higher level of unmet need for work than that of elderly males.

However, if the results based on the measure using the direct information on ability and desire to work in 1995 are considered, regardless of residence, elderly males displayed a higher level of unmet need for work than elderly women (5.4 percent vs. 3.6 percent for last week unmet need and 2.4 percent vs. 1.4 percent for last year unmet need). Differences in the results based on the measures using different types of information suggest that the measure using indirect information on ability and desire to work may overestimate the level of unmet need, and that direct information on the ability and the desire to work from the elderly respondents may be more appropriate.

Income Level and Trend

The most widely used index of economic status is income. The limitations of measures of income, as applied in the Thai context, have been extensively discussed elsewhere (Chayovan 1999). Despite existing problems, there are a number of distinct dimensions of income that can be operationalized and used in description and analysis. In addition to the absolute level of income of elderly, interest often centers on relative measures that permit comparison. For example, one may identify whether the respondent is in the top or bottom 10 or 20 percent of a distribution, or above or below the poverty line. In this paper the percentage of elderly with an income of less than 10,000 Baht per year is used as a rough indicator of the poverty line.

Questions about the total income of the respondents and their spouses (if any) during last year were asked in both the 1986 and 1995 surveys. In the 1986 survey, respondents were asked to report the absolute amount of their average monthly or yearly income during the year prior to the survey (Q438). In the 1995 SWET, respondents were first asked to give the absolute amount of their total last year income (K14). Those who failed to give the answer were asked to select a category of income groups read out to them by the interviewers. The procedure was an attempt to obtain an estimate of income level. It was found that non-response to the first question on income was high and the reading of income categories to respondents for them to choose from reduced the non-response from 34.5 percent to 11 percent (unweighted) out of the 4,486 total respondents (data not shown). The 1986 SECAPT, however, did not instruct the interviewers to read out the income categories for the respondents to choose from if they were unable to provide an answer on their absolute income. Thus non-response on the income question in SECAPT was higher than that of

Table 5. Mean, median and distribution of elderly according to income (in Baht) by type of respondent.

Income (in Baht)	Married Couples	Unmarried Males	Unmarried Females	Total
SECAPT (1986)[a]				
Mean	**16,302**	**9,795**	**6,866**	**12,200**
Median	**5,400**	**3,000**	**2,400**	**3,600**
(Median adjusted for inflation)	**8,802**	**4,490**	**3,592**	**5,388**
No Income	5.5	9.5	10.8	7.8
Less than 2,500	26.4	37.1	42.0	33.2
2,500 to 4,999	14.8	14.9	13.5	14.3
5,000 to 9,999	20.2	17.4	14.4	17.8
10,000 to 19,999	13.5	8.7	10.6	12.0
20,000 to 29,999	7.0	3.5	3.6	5.4
30,000 to 49,999	7.2	5.2	3.3	5.6
More than 50,000	5.3	3.8	1.8	3.9
Total	**100**	**100**	**100**	**100**
Unweighted N	1,460	285	1,009	2,754
Missing Income (%)	13.4	14.1	18.3	15.3
Unweighted N	1,686	333	1,233	3,252
SWET (1995)[b]				
Mean	**37,622**	**16,509**	**12,971**	**29,016**
Median	**15,000**	**5,000**	**3,750**	**10,000**
No Income	5.5	13.5	19.4	10.1
Less than 2,500	10.6	18.8	22.1	14.5
2,500 to 4,999	8.6	15.4	12.5	10.2
5,000 to 9,999	12.9	12.8	14.3	13.3
10,000 to 19,999	17.7	14.0	12.5	16.0
20,000 to 29,999	12.8	7.5	6.5	10.6
30,000 to 49,999	12.7	9.5	7.0	10.8
More than 50,000	19.2	8.6	5.8	14.6
Total	**100**	**100**	**100**	**100**
Unweighted N	2,069	370	1,554	3,993
Missing Income (%)	6.7	13.1	13.9	9.4
Unweighted N	2,231	428	1,827	4,486

Source: a) 1986 SECAPT; b) 1995 SWET.

SWET. The weighted proportion of respondents who did not have income information was 15.3 percent in 1986 and 9.4 percent in 1995 (Table 5). For both surveys, unmarried women display the highest proportion with missing information on income and married couples the lowest. Although the data from the two surveys are not directly comparable, they can be used to trace changes of the income position of the elderly.

Table 5 presents the amount and distribution of income of the elderly by marital status and gender for 1986 and 1995. Since married couples often pool income, the income of currently married respondents was the joint income with their spouse and their major source of income was also reported as a couple. Given that currently married elderly differ in many aspects from those who are never married, or widowed, or divorced or separated (i.e., have no spouse), the results presented in Table 5 are classified by gender and marital status. Based on the information on sex and marital status of the respondents, elderly are divided into three groups: currently married or married couples, unmarried males and unmarried females. Unmarried males and females include elderly who are never married, or widowed, or divorced or separated. It should be noted that the income of currently married elderly or married couples is always higher than that of unmarried males or females because the income of the previous group is the combined income of the elderly respondent and their spouse. Comparisons of income levels of the three groups need to take this into consideration.

Results in Table 5 reveal that in 1995, Thai elderly had an average annual income of 29,000 Baht and a median income of 10,000 Baht. More than one third had an income of less than 5000 Baht per year. About half had an income between 5,000–49,999 Baht. Only 15 percent had an income over 50,000 Baht. During 1986–995, median income of the elderly increased by almost three-fold, from 3,600 Baht to 10,000 Baht. After inflation is taken into consideration, however, the actual improvement in the income was found to be only about half of the original estimate. The median income has actually improved by 86 percent.

Although an increase in the proportion with 'no income' is observed, interpretations should be made with caution for several reasons. First, the results of these two surveys may not be directly comparable because of differences in the way the question was asked and recorded. Secondly, the 'no income' category is a unique group and does not necessarily imply an extremely poor economic condition. Some elderly may report no income because they were not working and did not consider income from children or interest from savings as part of their income. Other elderly may actually have no income but all their daily expenses are provided for by children or family members. Detailed examinations of some characteristics of 'no income' elderly revealed that all were able to provide information on the major sources of their income, and about 80 percent of them reported their children as the main source. About 9 out of 10 of these elderly lived with someone else; a majority had at least one asset and no debt; and 73 percent were satisfied with their financial situation (data not shown). Overall, their economic situation seems to be better than the average.

The improvement in income level differs between the three groups of respondents, namely currently married, unmarried males and unmarried females. Currently married elderly appeared to gain the most, while unmarried females the least. From 1986 to 1995, after taking into account inflation, the median income increased by 86 percent for married couples, 11 percent for unmarried males, and only 4 percent for unmarried females.

For all three groups, the income of the urban elderly is higher than that of rural elderly (Table 6). Unmarried men who are widowed are considerably poorer than those who are never married/divorced/separated. Among unmarried females, there is no difference in the income level by marital status.

Findings of this study suggest that although the income situation of Thai elderly may have improved during the past decade, many are still living in poverty, particularly unmarried females, and that overall their income levels are relatively low. The income of the elderly

Table 6. Median income and percentage with income of less than 10,000 Baht by selected characteristics, marital status and sex, 1995.

	Currently married		Unmarried male		Unmarried female	
	Median	% <10,000	Median	% <10,000	Median	% <10,000
Sex						
Male	15,000	34.7	–	–	–	–
Female	10,000	42.0	–	–	–	–
Residence						
Bangkok	40,000	18.6	15,000	43.6	15,000	41.0
Provincial urban	50,000	17.4	10,000	40.9	7,575	51.1
Rural	10,575	41.0	4,586	63.2	3,000	73.9
Marital status						
Never married/Div/Sep			10,000	48.2	3,750	68.8
Widowed			5,000	63.0	3,750	68.2

Source: 1995 SWET.

population is several times lower than the national average income. In 1995 the GNP per capita of Thailand was 72,956 Baht (US$ 2806, based on the exchange rate of 26 Baht/US$ 1 in 1995). The median income of the elderly is at or below the poverty line as defined by the National Economic and Social Development Board (NESDB)[a]. NESDB's findings and the results of this study indicate that not only is the poverty line of women lower, but their actual income is also lower than that of men, reflecting a less favourable economic condition for women compared to men. This is especially true in the case of unmarried women.

Debt and Savings

The economic condition of the elderly depends on more than just their current work status and income level. Further insights can be provided from information on the prevalence of owing debts and having savings or bank account or stocks. Unfortunately, this information was collected in the 1995 survey only; hence it is not possible to examine the time trends from this aspect. The extent of being in debt was measured by owing any debts, and by owing a significantly large amount, defined arbitrarily as 25,000 Baht or more.

Data in Table 7 indicate that about one-fourth of Thai elderly are in debt and 11 percent report a debt of at least 25,000 Baht. There are differences in the level and the amount of debt by gender and place of residence. Elderly men display a higher proportion of debt and urban residents are less likely than rural residents to report being in debt. As for the amount of debt, the same patterns of gender and residential differentials exist.

---

[a] A study conducted by NESDB (1998) reported that the minimum need of an individual vaires by age, sex, region and area of residence. They also found that the minimum needs of persons aged 60 and over ranges from 600 Baht per month for women in the sanitary and rural areas of the Northern region to 982 Baht per month for men in the municipal areas of the Southern region (or between 7,200 to 11,784 Baht per year).

Table 7. Percentage of elderly who are in debt, who have debts of 25,000 Baht or more, and percentage who have savings, by sex and residence, 1995.

		Total		Urban		Rural	
	Total	**Male**	**Female**	**Male**	**Female**	**Male**	**Female**
% In debt	24.3	30.9	18.7	17.5	10.1	33.5	20.8
% In debt 25,000 baht +	11.0	15.1	7.7	11.8	5.1	15.7	8.3
% Having savings	31.9	35.2	29.1	48.8	37.1	32.6	27.2

Source: 1995 SWET.

Savings and wealth accumulated over a lifetime has a bearing on current work status or retirement decision. Wealth and income are different economic concepts. In general, elderly tend to have more wealth but lower income than the average adults. A majority (80 percent) of Thai elderly own the house they live in. A higher proportion of rural elderly than urban elderly own the house. This has not changed during the period 1986 to 1995 (Chayovan et al. 1988; Chayovan and Knodel 1997).

When respondents were asked about assets, savings, bank accounts or stocks were combined in the same question. Presumably most who answered affirmatively were referring to having savings in some form including a bank account or stocks. Thus, this question refers to savings. Unfortunately, the amount and duration of the savings were not asked. Results in Table 7 show that about only one-third of elderly report that they have savings. Men are more likely to have savings than are women in both urban and rural areas. Rural men and women, however, are less likely to have savings than their urban counterparts. This is consistent with urban-rural differences in the income level.

Sufficiency of Income and Satisfaction with Financial Situation

Satisfaction with and sufficiency of income can be used as a proxy for, or subjective assessment of, the overall picture of economic well-being. To some extent, the two may reflect both a person's income and expense levels.

The two surveys asked the questions differently. The 1986 survey asked respondents who had income to indicate if their total income was more than sufficient, sufficient, or insufficient for their own expenses (Q439), while SWET asked respondents to indicate their general level of satisfaction with their current financial situation. They were given three choices: very satisfied, satisfied, and not satisfied. Results presented in Table 8 suggest an improvement in the economic condition of the Thai elderly. The proportion who reported dissatisfaction with their financial situation in 1995 is significantly lower than the proportion who stated that their income was insufficient in 1986. Also, findings from the 1994 survey of elderly in Thailand which asked a similar question on the sufficiency of income to the 1986 SECAPT show that the percentage who indicated insufficiency of income declined from 51 percent in 1986 to 35 percent in 1994 (NSO 1995). This is consistent with the rise in their income level during 1986–1995. Despite improvement in the economic situation,

Table 8. Percent distribution of elderly according to the sufficiency of income by sex and residence, Thailand, 1986.

		Urban		Rural	
	Total	Male	Female	Male	Female
Sufficiency of income (1986)[a]					
More than sufficient	5.5	9.4	12.7	5.0	3.8
Sufficient	43.9	52.5	56.8	39.2	43.3
Not sufficient	50.5	38.2	30.5	55.8	52.9
Total	**100**	**100**	**100**	**100**	**100**
Satisfaction with financial situation (1995)[b]					
Very satisfied	9.1	13.9	14.4	8.4	7.7
Satisfied	61.7	65.7	69.0	57.0	63.5
Not satisfied	29.1	20.4	16.6	34.6	28.8
Total	**100**	**100**	**100**	**100**	**100**

Source: a) 1986 SECAPT; b) 1995 SWET.

the proportions of elderly who had insufficient income and who were not satisfied with the financial situation were still quite substantial (51 percent and 29 percent respectively).

For both urban and rural areas, males are more likely to report the insufficiency of their income or express dissatisfaction with their financial situation than females. Regardless of urban–rural differences, the proportions of elderly who had insufficient income and who were dissatisfied with their financial situation among rural males and females are higher than urban males and females.

Findings in the previous section indicate that Thai elderly are relatively poor, but almost half (49 percent) of elderly in 1986 stated that their income was sufficient, with about 6 percent reporting their income was more than sufficient. In addition, a majority (71 percent) of elderly in the 1995 survey were satisfied with their financial situation, with 9.1 percent reporting that they were very satisfied, and only 29 percent were dissatisfied. The high concentration of responses in the 'sufficient' and 'satisfied' categories may be partly a reflection of a characteristic of Thai people. Thais are known to be flexible and value self-control, polite and avoid conflict in interpersonal interaction (Komin 1991). These values may influence the elderly to respond to questions either positively or neutrally. In addition, a Buddhist teaching to practise the moderate path may influence a 'sufficiency attitude' amongst the Thai. Thus, although they may be poor by international standards they are likely to adjust to and be satisfied with what they have.

The level of satisfaction with one's economic situation in large part depends on the level of income and financial security. Income level has been shown to be positively associated with the level of satisfaction with one's financial situation, and that currently married elderly are least satisfied, while unmarried females are the most satisfied with their economic situation (Chayovan 1999). The unmarried females who are considered the most economically

Table 9. Percent distribution of elderly according to comparison of the financial situation now with when aged 40–50, by sex and residence, Thailand 1986.

Compare financial situation now with when aged 40–50	Total	Urban		Rural	
		Male	Female	Male	Female
Better	23.3	23.9	30.1	22.3	22.5
Same	14.9	15.3	18.6	13.2	15.3
Worse	61.8	60.9	51.3	64.6	62.2
Total	100	100	100	100	100

Source: 1986 SECAPT.

vulnerable group expressed the highest level of satisfaction with their financial situation. This probably points to the need for better measurements of vulnerability. It is also possible that men are expected to support their family and thus feel the pressure of earning enough to provide for the family member, while women may be more adaptive to difficult situations than men. The high level of dissatisfaction among currently married elderly suggests that absolute income alone is not the best measure of economic well-being.

It has been documented that an individual's financial situation is likely to deteriorate as age increases (Kinsella and Gist 1995). SECAPT asked the respondents to assess their current financial situation when they were 40–50 years old and results are shown in Table 9. Approximately 62 percent of elderly reported that their current financial situation was worse than when they were younger while 23 percent reported that it is better. As expected, rural elderly more than urban elderly stated that their current financial situation was worse than when they were younger. Higher percentages of both rural and urban men than women reported that they were in a worse financial situation than when they were younger.

Main Sources of Income

Examining the main sources of income for older persons helps explain the continuance of poverty in old age and the existence of marked income inequalities among Thai elderly. The source of income is an important determinant of both the amount and regularity of income and thus is influential for the economic well-being of the elderly. The more sources of income the higher the level of income expected. Sources of income for living expenses also reflect on both how the elderly live their lives and who provides the main support for them.

Information on the main source of income by gender and urban-rural areas for 1986 and 1995 is presented in Table 10. It should be noted that the questions asked in the two surveys were slightly different. The 1986 survey asked respondents about major sources of income or wealth for their living expenses, and if more than one source was given they were asked to give the most important source (Q436 and Q437). The 1995 survey only asked the most important source of income (K19). Thus, time trends on the income sources can be examined only for the most important source.

Table 10. Percent distribution of elderly according to main source of income by residence, 1986 and 1995.

	1986[a]			1995[b]		
Main source of income	**Total**	**Urban**	**Rural**	**Total**	**Urban**	**Rural**
No income	6.7	4.9	7.0	–	–	–
Work	28.4	17.3	30.7	29.5	21.7	31.3
Pension	2.3	9.2	0.9	2.4	5.1	1.8
Savings	8.3	9.2	8.2	4.4	6.5	4.0
Children	47.5	51.7	46.7	49.3	50.8	49.0
Spouse	2.0	2.8	1.8	8.3	10.6	7.8
Relatives/other	4.9	5.0	4.8	6.0	5.3	6.2
Total	**100**	**100**	**100**	**100**	**100**	**100**

Sources: a) 1986 SECAPT; b) 1995 SWET.

Data in Table 10 suggest that the main sources of income for the elderly in Thailand remain almost unchanged during 1986 and 1995. There are two predominant sources of income for Thai elderly, namely children and work. Other sources such as pension, savings and other relatives are far less significant. In 1986, almost half of the respondents reported that their main source of income was from their children, while 28 percent stated that it was from their own or their spouse's work. About 8 percent had income from savings and only 2 percent from a pension. In 1995, children still remain the predominant source of income, followed by work. The third most important source in 1986 was savings, while in 1995 it was the spouse. The proportions who report a pension as a main source remain the same at 2 percent, suggesting no change in the coverage of the formal support systems during 1986–1995. Although a small percentage of elderly received their income from a pension, these individuals had the highest amount of income due to the regularity of income from a pension (Chayovan 1999).

Gender and residential differences in the main sources of income are evident. Men are more likely than women to receive income from work and pension, but less likely to receive income from their children and spouse. The proportion receiving income from a pension among men is considerably higher than that for women, a reflection of higher education levels among men and their ability to get employment in government services.

Elderly living in rural areas have a higher proportion that receive income from work rather than from their children compared to those living urban areas. Urban residents are more likely to have income from savings and a pension than those living in rural areas, a reflection of their better education, and access to financial security. Other relatives/other is also an important source of income for small minorities of both urban and rural elderly.

To project sources of income for future elderly, SWET asked the near elderly, defined as those who were 50–59 years of age, what were their expected main sources of income for old age. Although multiple sources are possible, respondents were asked to give only one

Table 11. Percent distribution of persons aged 50–59 according to expected source of income for old age, by sex and residence, 1995.

Expected source of income for old age	Total	Total		Urban		Rural	
		Male	Female	Male	Female	Male	Female
Children	22.0	21.3	22.6	22.6	22.1	21.1	22.7
Son	10.5	10.0	11.0	6.2	11.6	10.8	10.8
Daughter	20.4	16.2	23.8	9.7	15.3	17.6	25.9
Work	15.6	19.4	12.4	21.0	15.6	19.1	11.6
Savings/interests	12.0	14.6	9.8	16.0	14.2	14.3	8.7
Pension	2.3	4.0	0.8	9.7	2.1	2.8	0.5
Spouse/relatives/other	2.6	1.8	3.2	2.6	5.7	1.6	2.6
Never thought	14.6	12.6	16.3	12.2	13.5	12.7	17.1
Total	**100**	**100**	**100**	**100**	**100**	**100**	**100**

Source: 1995 SWET.

source that they thought would be the major one. Results in Table 11 confirm that informal support systems, particularly children (53 percent) will continue to be the primary provider of old age support, followed by work (16 percent), savings/interest (12 percent), spouse and other relatives (3 percent) and pension (2 percent). Among those who specified the gender of the child from whom they expected support, the proportion mentioning daughters is considerably higher than the proportion mentioning sons (20 percent vs. 10 percent). This is true for both men and women regardless of residence. These Thai results are consistent with findings in the Philippines where bilateral family systems are prevalent and where married daughters are more likely than sons to co-reside with older parents (Ofstedal, Knodel and Chayovan 1999).

Expected main sources of old age income varies by sex and residence. For both rural and urban areas, men are more likely than women to expect income from work, savings and pension, but less likely to expect their children and spouse/relatives/other to be their main sources of support.

About 15 percent of the near elderly stated that they never thought of sources of support for old age. Information on the characteristics of this group may provide further insight into the potential problems of the elderly. The results of changes in the main sources of elderly's income and the expected sources of income of the near elderly suggest that informal support systems are key to the survival of the elderly in Thailand.

Preparation for Old Age

Evidence provided thus far indicates that the informal systems of support will continue to be a primary source of old age security for the elderly in Thailand. The impact of fertility decline on the familial support system in the case of Thailand was found to be relatively moderate (Knodel, Chayovan and Siriboon 1992). Although exchanges of visits are associated with the location of children and elderly, Thai elderly receive frequent

Table 12. Percent distribution of elderly according to type of preparation for old age, by sex and residence, 1986.

		Urban		Rural	
	Total	Male	Female	Male	Female
Not prepared	56.5	58.8	60.6	51.6	58.7
Savings	20.5	25.7	23.4	21.3	18.5
House and or land	11.2	4.8	3.7	16.7	9.9
Other	10.0	9.7	9.3	9.0	10.8
Don't know the type	1.9	0.9	3.0	1.4	2.1
Total	100	100	100	100	100

Source: 1986 SECAPT.

visits and material support from non co-resident children (Chayovan and Knodel 1997). However, to ensure the quality of life of the elderly, efforts should be made so that not only the family, the community, the government and private sectors take responsibility, but that the individuals also prepare themselves for old age. Different forms of preparation could be made including accumulating wealth.

The 1986 SECAPT asked respondents if they think preparation for old age is important and whether they have ever made any preparations for old age (Q710 and Q711). Respondents who said they have made preparations for old age were asked about the types of preparation and the most important preparation they have made (Q712 and Q713). Although multiple answers on the types of preparation were possible, only the most important type of preparation was coded.

Results from the 1986 SECAPT reveal that although a majority of the elderly viewed the preparation for old age as an important issue, more than half did not make any specific preparation for this stage in their life (Chayovan 1992). Differentials by gender and residence are given in Table 12. For both urban and rural areas, female elderly are less likely to make any preparation than male elderly. The higher proportion of elderly who made preparation for old age among males may partly be due to their sense of responsibility of being the main provider for their families. They may view the preparation for old age as a form of security for themselves and their families. It is observed that the proportion of elderly who did not make any preparations is higher for both elderly men and women in urban areas compared to their rural counterparts. Urban elderly may have better access to the formal support system and thus are less motivated to make personal preparations.

Financial issues appear to be of most concern among the elderly. About one-fifth of elderly reported that savings was the most important type of preparation they made, followed by having assets in the form of a house or land. Urban elderly tend to prepare in the form of savings while rural elderly are more likely to prepare in the form of assets. The higher house and land prices in urban areas may prevent urban elderly from owning them.

Table 13. Percentage of persons aged 50–59 who have prepared each type and percentage who have savings, by sex and residence, 1995.

		Total		Urban		Rural	
Type of preparation made	**Total**	**Male**	**Female**	**Male**	**Female**	**Male**	**Female**
Savings	37.3	42.0	33.4	46.2	42.1	41.1	31.3
Living with whom	41.6	41.0	42.1	37.1	40.1	41.8	42.6
How to spend free time	36.1	39.4	33.4	37.1	38.2	39.9	32.2
Property transfer	15.3	15.7	14.9	9.1	11.0	17.1	15.9
Exercise	47.5	54.3	41.8	59.8	45.9	53.1	40.8
Health check up	30.9	30.0	31.7	37.3	42.0	28.5	29.2
Become more religious	66.5	63.5	69.0	58.8	65.7	64.5	69.8
% who have savings	39.3	44.1	35.5	45.4	42.8	43.8	33.7

Source: 1995 SWET.

Although to some extent the SECAPT results may reflect the actual situation, they may also be affected by the nature of the question asked. SECAPT questions were retrospective and open-ended in nature. This may make it difficult for the respondents to answer. In SWET a different approach was taken with regard to questions related to preparation for old age. Questions were directed towards the near elderly respondents, defined as those who are 50–59 years of age. They were asked whether they thought about and whether they actually made any specific types of preparation for old age, namely savings, planning living arrangements, making plans to spend free time, handing over inheritance, doing physical exercise, having regular health checkups, and turning closer to religion. Each possible preparation was read to the respondent. It is noted that the percentage of the near elderly who have savings is almost the same as the percentage reporting savings as a type of old age preparation regardless of sex and rural–urban residence. This finding suggests that the percentage having savings may also be used as an indicator of financial security for old age.

Comparing the data in Tables 12 and 13 suggests that although preparation for old age among the near elderly is still moderate, it may gain more attention from the future elderly. The most common preparation made among the future Thai elderly is turning closer to religion (67 percent), doing exercise (48 percent), planning living arrangements (42 percent), saving money (37 percent), making plans to spend free time (36 percent), having regular health checkups (31 percent) and handing over inheritance (15 percent). These results suggest that future Thai elderly have a tendency to prepare for psychological rather than economic and health aspects. Economic hardships during their working life may prevent them from being able to save or accumulate wealth. For both urban and rural areas, men are more likely to make economic and health preparations while women tend to turn closer to religion. Preparations relating to financial and health matters are made more by urban rather than rural near elderly of both sexes. Near elderly in rural areas are more likely than their urban counterparts to prepare by turning closer to religion and by passing on inheritance.

Formal Financial Support Systems for Old Age in Thailand

Besides the free medical care for all elderly, a program started in 1989, the present programs to provide financial old age security in Thailand are very limited in scope and magnitude. Pensions are the financial lifeline of the elderly in most industrial countries. In Thailand, however, public pension systems generally cover only a small proportion of the labour force. Primarily only government and state enterprise retirees who have provided at least 25 years of service receive a pension. About 2 percent of the elderly are in this category and the coverage is not expected to expand in the near future, especially with the present policy of downsizing government organizations. The projection of the increase in the amount of pensions the government must pay for future retirees has prompted the government to set up the "Pension Fund for Civil Servants" Scheme in 1996. Under this new scheme, the insurers have to contribute 3 percent of their monthly salary to the fund. The benefits under this scheme are, by and large, similar to that of the old system. As of December 1999, there were 1.12 million government employees insured in this program (Kesornsutjarit 2000).

The public social security system in Thailand is still very young, established only in 1990. As of January, 2000, about 5.18 million workers or about 17 percent of the total labour force were insured by this system (Kesornsutjarit 2000). Under this scheme, the insurers have to contribute about 3 percent of their monthly salary for a minimum of 15 years continuously and must be over 55 years of age to be eligible for the old age benefits. The retirees will receive an average of about 20 percent of their average last five-years salary. Moreover, it will not be before the year 2013 that the old age benefits will be paid for the first time.

In addition to the above mentioned two systems of old age security, the government, through the Department of Public Welfare, initiated a financial assistance program for the indigent elderly in 1993. The program provides a monthly allowance of 200 Baht for the poor elderly who have no one to support them. Approximately 350,000 elderly nationwide (or about 7 percent of all elderly) receive this monthly allowance.

Conclusion and Discussion

Thailand has been relieved from the pressure of population increase, but is now facing a new pressure from population ageing as a result of changes in population structure. Population projections have demonstrated a transformation of the population age structure and signalled the acceleration of the tempo of population ageing Thailand will experience. The old-age dependency ratio will accelerate in the next decade. While the conventional dependency ratio, to some extent, may overestimate the burden of older people that the working population has to shoulder, it provides an approximate assessment of the potential negative socioeconomic consequences. The rapidity of ageing of the population in Thailand is likely to have an unusual effect on Thailand's economy and society because older people consume relatively high proportions of the national budget for health, financial assistance and social services and welfare. Experiences of the industrialized countries indicate that ageing has brought about serious public financial problems as a result of high government expenditures, especially on pensions and medical care because of the unprecedented increase in the number of older persons. One way of coping with this unprecedented ageing in Thai society is to consider policies and programs that would alleviate the poverty, inequality and age pressure, and ensure the quality of life of the old.

Although the future of older Thais will differ from those of today, one aspect that is likely to remain is their poverty and inequality. Overall, a fairly consistent picture emerges from this review of changes in the economic situation of Thai elderly with respect to gender and urban–rural differences. By and large, Thai elderly people are trapped in poverty. Poverty and inequality in old age are primarily a function of access to resources over the earlier stages of the life cycle. Occupational status in the earlier stage of life determines salaries or wage levels which in turn determine the opportunities to save and invest in property and other possessions for old-age security. Access to pensions also rest on employment status during earlier working years. Survey results show that although employment levels of the elderly may have increased during 1986–1995, a high proportion of them do not continue working after retirement age because of health-related reasons. Elderly who do not work generally have lower income levels than those who do. Thai elderly have to depend heavily on their families who may also struggle to make ends meet. A substantial proportion of elderly report either having insufficient income or are dissatisfied with their financial situation. In addition, most of the present and future elderly in Thailand have no savings and have not made any preparation for old age. Gender and rural–urban inequalities in the status and accessibility of economic resources still persist over time. Elderly women and rural elderly are generally worse off economically than elderly men and urban dwellers.

It is unfortunate that Thailand is experiencing an ageing of its population at the same time that the country is facing an economic down turn. The financial crisis is perhaps the primary current constraint or drawback on service and program development for the elderly. The impact may also be long-term because after more than five years of economic slump there is no strong sign of recovery. This may be the time that the country has to start seriously re-evaluating the nature of its social, health and financial programs for the elderly. A new model of public–private sector participation and cost sharing may be brought in. For the elderly, the crisis has been of more relevance in its immediate impact on their families and is in urgent need for the assessment of its impacts on the well-being of the elderly.

Given the rapid rate at which age structures are changing, both the public and private sectors should intensify their efforts to plan for meeting the needs of a rapidly increasing number of elderly persons. Clearly, it is unrealistic to expect the government of a developing country such as Thailand to provide wide services for all elderly. Indeed, in recognition of this, the government hopes to rely on the continuation of familial support for meeting the bulk of the need of the majority of elderly persons. As the economy expands and the country becomes more urbanized, informal support systems may not prove to be as widely available. Continued reliance on family support could also put great strain on family members, particularly women.

In addition to taking measures to facilitate and supplement family efforts to support their elderly, the government must face this new challenge by reforming its existing systems and initiate new strategies or develop additional old-age schemes to complement the existing schemes. The draft of the second long-term plan for the elderly, 2001–2021, currently being prepared, puts an emphasis on lifelong preparation to ensure old age security and well-being. Elderly are considered an integral part of society. Ageing issues are the responsibilities of everyone: the family, the community, the government and the elderly themselves. With limited resources, the government cannot alone support the growing number of elderly.

Programs and strategies should be designed not only to involve all levels of governmental and non-governmental bodies (e.g., private firms and unions), but also to encourage self-reliance in terms of income, employment and health care. As Buddhism teaches that "self-reliance is best", financial security during old age cannot be solely dependent on welfare or government support: individuals should help themselves through employment, personal savings and other preparation. The government may facilitate and encourage individual savings for old age by exempting taxes on such accounts. Expansion of the social insurance scheme to cover self-employed workers and farmers must be immediately implemented. A new provident fund may be set up for home makers who have no access to an occupational pension.

References

Chayovan, N. (1992), "Preparation for old age among the Thai adults and elderly." (in Thai). A paper presented at the *Annual Meeting of the Thai Population Association*, pp 163–194.

Chayovan, N. (1999), "Economic well-being of Thai elderly: gender differences." A paper presented at the *International Conference on Women's Health—The Nations Gain: With a Special Focus on Older Women in Asia, Organized by the KK Women's and Children's Hospital and the Tsao Foundation, Singapore*, July 5–7.

Chayovan, N. and Knodel, J. (1997), *A Report on the Survey of the Welfare of the Elderly in Thailand*, Publication No. 248/97. Institute of Population Studies, Chulalongkorn University: Bangkok.

Chayovan, N., Wongsith, M. and Chanpen S. (1988), *Socio-Economic Consequences of the Ageing of the Population in Thailand: Survey Findings*, Publication No. 161/88. Institute of Population Studies, Chulalongkorn University: Bangkok.

Kesornsutjarit, J. (2000), "Social security system in Thailand: past experiences, obstacles and ways to reform." Paper prepared for the *Third APEC Regional Forum on Pension Fund Reform, Bangkok, Mar. 30–31.*

Kinsella, K. and Gist, Y.J. (1995), *Older Workers, Retirement, and Pensions: A Comparative International Chartbook*, U.S. Department of Commerce, Bureau of the Census.

Knodel, J., Chayovan, N. and Siriwan, S. (1992), "The impact of fertility decline on familial support for elderly: an illustration from Thailand." *Population and Development Review*, 18:79–102.

Komin, S. (1991), *Psychology of the Thai people: Values and Behavioral Patterns*, Research Center, National Institute of Development Administration (NIDA): Bangkok.

National Economic and Social Development Board (NEDSB), Development Evaluation Division. (1998), "A Newsletter: Indicators of Well-Being and Policy Analysis." *Poverty Line and Use*, 2(2): 1–12.

National Statistical Office (NSO), Office of the Prime Minister. (1988), *A Report on the Labour Force Survey (3rd Round), 1985, Bangkok.*

National Statistical Office (NSO), Office of the Prime Minister. (1996), *A Report on the Labour Force Survey (3rd Round), 1995, Bangkok.*

National Statistical Office, Office of the Prime Minister. (1997), *A Report on the Survey of Population Change, 1985–86, Bangkok.*

Ofstedal, M/B., Knodel, J. and Chayovan, N. (1999), "Intergenerational support and gender: a comparison of four Asian countries." *Research Report*, No. 99-54, Population Studies Center, University of Michigan.

CHAPTER 10. CHANGING FAMILY STRUCTURE IN TURKEY, 1968–1998

TURGAY ÜNALAN
Institute of Population Studies
Hacettepe University, Turkey

1. Introduction

All over the world, changing patterns of fertility and migration over the last 30 years have significantly affected people's lives. In many parts of the world, growing numbers of people reaching to older ages, rapidly declining levels of fertility, increasing mobility of people both nationally and internationally, combined with rapid social and economic changes resulted in adaptations of the family, the basic unit of all societies, to these new challenges and opportunities (Unfpa, 1999). This paper examines the change in the household or family composition of the Turkish society during the period 1968–1998 and tries to explore how the age composition and the type of families people are living in have been affected by the demographic changes and the changes in the household composition.

1.1. THE DEMOGRAPHIC TRANSITION IN TURKEY

After its foundation in 1923, the new Turkish Republic experienced substantial changes in its demographic structure. However, this renewal process brought a lot of changes in demographic characteristics. There is now a completely new balance among age groups and between the two sexes. The geographical distribution of population, the density of habitation, and population size all changed fundamentally. Turkey is now a much more urbanised country, health is better and people live longer. Families no longer have many children, though individual variation still exists. All of these changes had and continue to have profound effects on the social, political, and economic life of the country (SIS, 1995).

The demographic transition of the country has been accomplished in three stages. The first stage of transition can be dated from 1923 to 1955. Rates of population growth were below 2 percent per year until the late 1940s. Both the First World War and the War of Independence during 1919–1923 caused massive population losses in the country. After 1923, peace and the recovery of normal life led to a steady decline of death rates, except for a brief reversal during the Second World War. However, fertility increased significantly from around 5.5 children to 7.0 children during this period. During the first decades of the Republic, in order to rebuild family and social life and to overcome the shortages of labour force, a pro-natalist policy encouraging high birth rates was adopted and remained valid until 1965. With decreasing death rates and increasing birth rates, the population growth

S. Tuljapurkar et al. (eds.), Population, Resources and Development, 181–201.

rate increased rapidly: the population almost doubled between 1923 and 1955, increasing from 13 to 24 million.

The second period of Turkey's demographic transition can be dated from 1955 to 1985. This period starts with the population growth rate of 2.8 percent per year. During the 1950s, fertility began to decline and it did not reverse. However, the pace of decline was not fast enough to catch up immediately with the previous decline in death rates, so the population continued to grow (SIS, 1995). Between 1955 and 1985, the population doubled again, from 24 to 51 million.

During the 1980s, Turkey entered the third phase. The decline in the rate of population growth was definite and irreversible. The population growth rate decreased from its highest level of 2.8 percent per year to 2.2 percent during the five-year period, 1985–1990, in spite of additions due to immigration. By 1990, the rate of population growth was 2.2 percent per year. The third stage of demographic transition is considered to be complete when two things happen: 1) fertility should fall to a level where births approximately replace the parent generation, but not more than that; and 2) population growth should stop (SIS, 1995). Between 1985 and 1997, the population increased from 51 million to 63 million. Based on the projections assuming that the total fertility rate will decline to replacement level by the period 2000–2005, the population of Turkey is expected to reach 76 million in 2010 and 88 million in 2025 (Ergöçmen et al., 1995).

1.2. POPULATION AND POPULATION GROWTH RATES

Turkey's population was 13.6 million in 1927 according to the first census that was held four years after establishment of the Republic in 1923. According to the result of the latest population count in 1997, Turkey has a population of 62.9 million. Population growth rates have continuously fluctuated since the first census. The annual growth rate is estimated to be 1.5 percent, down from 2.5 percent in the 1970s and 2.8 percent in the 1960s (Table 1). The total population is projected to reach 75 million in 2010, and 86 million in 2025, before stabilising at just less than 100 million.

The Turkish government supported family planning programs as early as 1965, seeing rapid population growth as an obstacle to economic development. In response to high fertility rates and rapid population growth in the 1950s and 1960s, the Turkish government encouraged both temporary and permanent emigration of its population. Nowadays, an estimated 3 million Turks are living abroad. More than 90 percent of them live in Europe. During the last decade, Turkey transformed itself from a country of emigration to a country of immigration (Iranians, Bulgarian, Ethnic Turks), of asylum (Bosnians, Iraqis), of transit (especially Ghana, Nigeria), and, still, of decreasing emigration.

1.3. AGE COMPOSITION OF POPULATION

Turkey has a young population as a result of high fertility and growth rates. According to the 1997 population count, those who are under the age of 15 constitute 31 percent of the total population (Table 2). Although elders who are above 65 have a small proportion of

Table 1. Population and growth rates in Turkey, 1927–1997.

	Population	Growth rate %
1927	13 648 270	–
1935	16 158 018	2.11
1940	17 820 950	1.96
1945	18 790 174	1.06
1950	20 947 188	2.17
1955	24 064 763	2.78
1960	27 754 820	2.85
1965	31 391 421	2.46
1970	35 605 176	2.52
1975	40 347 719	2.50
1980	44 736 957	2.07
1985	50 664 458	2.49
1990	56 473 035	2.17
1997	62 865 574	1.51

Source: SIS (1993) and SIS (1999).

Table 2. Percentage distribution of population by age groups, Turkey, 1935–1997.

	0–14	15–64	65+	Median age
1935	43.1	53.0	3.9	NA
1940	42.1	54.4	3.5	NA
1945	39.5	57.2	3.3	NA
1950	38.3	58.4	3.3	20.1
1955	39.4	57.1	3.4	20.4
1960	41.2	55.2	3.5	20.3
1965	41.9	54.1	4.0	19.0
1970	41.8	53.8	4.4	19.1
1975	40.6	54.8	4.6	19.3
1980	39.1	56.1	4.7	19.8
1985	37.6	58.2	4.2	21.4
1990	35.0	60.7	4.3	22.2
1997	31.0	63.2	5.8	23.8*
2000	29.6	64.9	5.5	25.6
2005	27.6	66.6	5.8	27.6
2025	21.9	68.4	9.7	34.2
2050	19.4	61.9	18.7	39.1

Source: Shorter and Macura (1982), SIS (1993), Unpublished results from the 1997 Population Count, SIS (1995), and United Nations (1999).
* Estimate for the year 1995.
NA: Not Available.

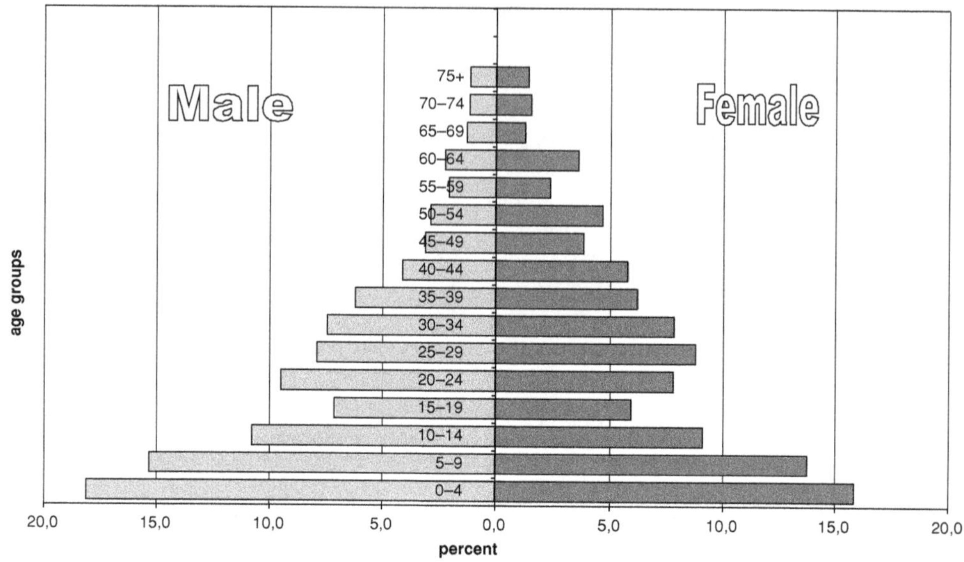

Figure 1. 1935 Age pyramid.

only 5.8 percent in the total, it is expected that this percentage will increase rapidly in the next few decades as a result of declining fertility and mortality levels.

The overall trend in the age structure of Turkish population is towards an increasing proportion of elderly and a decreasing proportion of young people. With regard to absolute numbers, the young population is below 20 million and its increase has slowed down while the elderly population continues to increase rapidly (from around 2.1 million in 1980 to 2.4 million in 1990 and to 3.6 million in 1997). With decreasing mortality and fertility levels, the proportion of older population is expected to rise more rapidly than any other age group in Turkey. Population projections for the year 2000 show the proportion of elderly population as 5.5 and increasing to 9.7 by the year 2025.

The progress of Turkey's population during the Republic period can be best viewed with the help of population pyramids (Figures 1 to 6). Tabulation for five-year age groups became available starting from the 1935 population census. The age pyramids presented for various time intervals reflect the transition of the country from a high fertility one during the first half of the century to a medium fertility one in the second half of the century. Projections for the first half of the new millennium assume low fertility in Turkey. The present-day age pyramid (Figure 4) reflects the effects of high fertility in the past; and evidence of recent fertility declines is reflected in the smaller proportions of children under age 15.

The total dependency ratios over time display a steady decline during the last decade (Table 3). According to the latest data, total dependency ratio has declined to 58, meaning that there are 58 dependent persons to each 100 persons who are of working age. The large

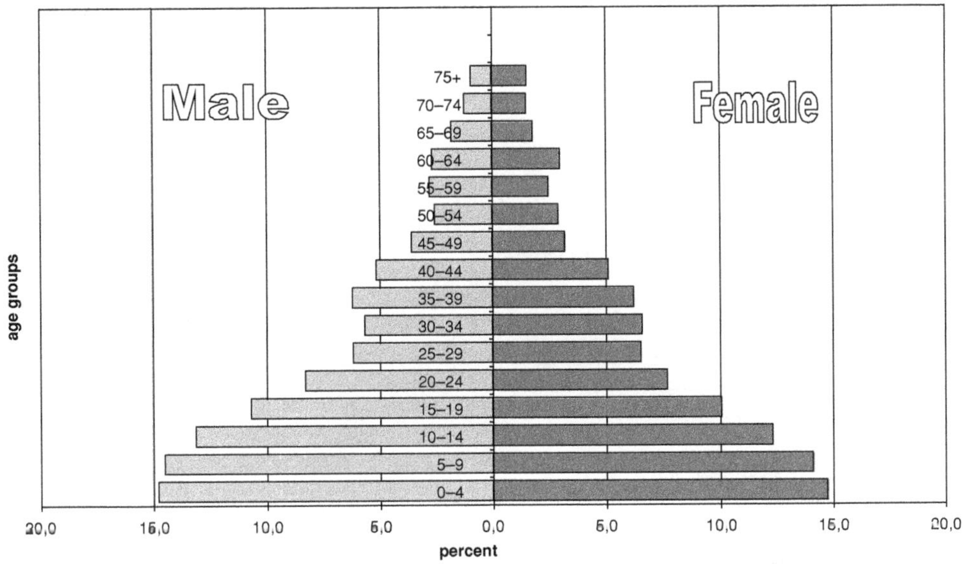

Figure 2. 1970 Age pyramid.

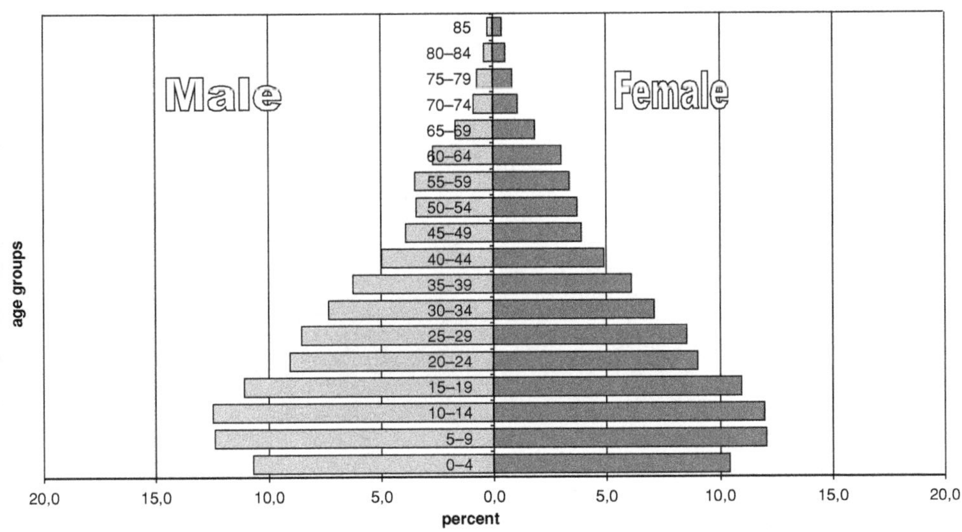

Figure 3. 1990 Age pyramid.

share of the children among dependent persons seems to be declining over time while that of the elderly increases. Even though there is a declining trend, relatively high levels of children point to the necessity of enlarging employment opportunities and to a need for creating education and health infrastructures on a greater scale.

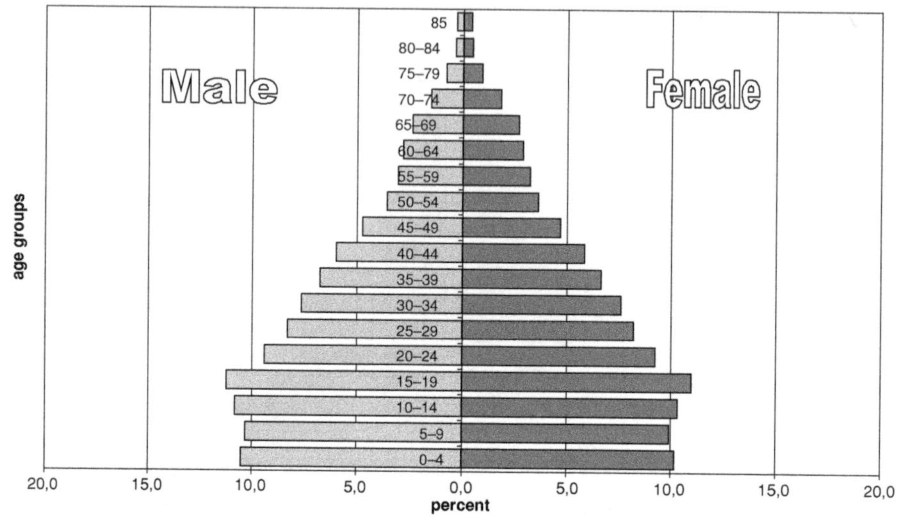

Figure 4. 1997 Age pyramid.

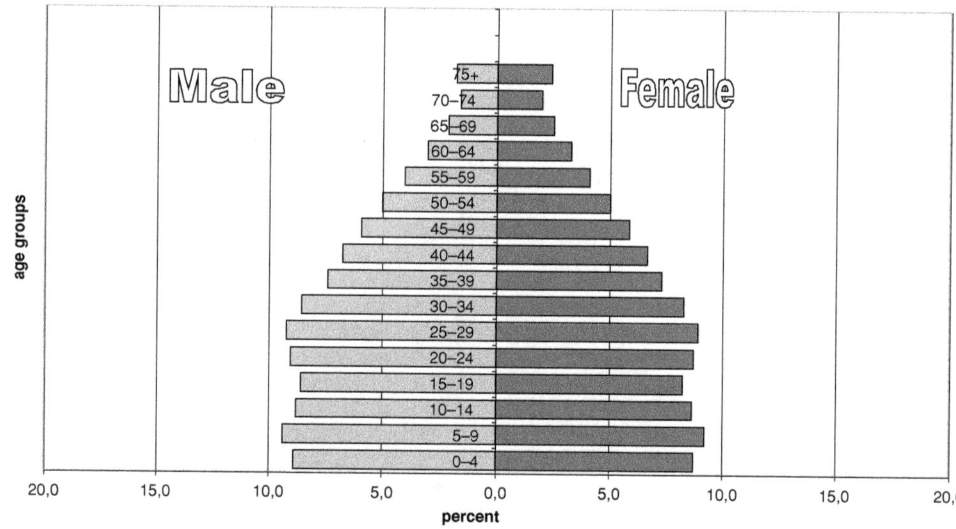

Figure 5. 2010 Age pyramid.

The high ratio of children to working age adults reached its peak around the year 1965 and then started to decrease. Although the elderly population comprises a relatively small percentage of the total population, their numbers are growing.

Another measure of demographic ageing is median age. In 2000, the median age of the population is estimated as 25.6 years, up from 20.1 years in 1950. It is projected to reach 34.2 years by the year 2025 and to 39.1 years by 2050. Within the same period, the medium

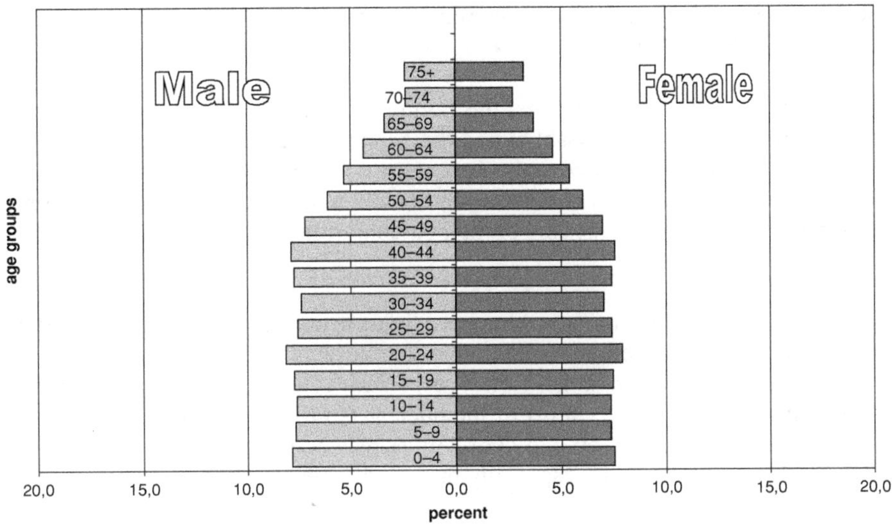

Figure 6. 2025 Age pyramid.

fertility scenario of the United Nations indicates that the median age of the world population will rise from 25.6 years in 2000 to 39.1 years in 2050.

1.4. THE FAMILY IN TURKEY

The family has always been the centre of social life in Turkey. Contrary to expectations for a Muslim country located close to the Middle Eastern world, the typical family of Ottoman

Table 3. Age dependency ratios, Turkey, 1935–1997.

	Total	65 years and over	0–14 years
1935	82.9	7.1	75.8
1940	78.9	6.5	72.4
1945	75.1	5.9	69.2
1950	71.3	5.7	65.7
1955	75.0	6.0	69.0
1960	81.1	6.4	74.7
1965	84.9	7.3	77.6
1970	85.9	8.2	77.7
1975	82.3	8.4	73.9
1980	78.1	8.5	69.7
1985	71.8	7.2	64.6
1990	64.7	7.1	57.6
1997	58.2	9.1	49.1

Source: SIS (1993) and unpublished results from the 1997 Population Count.

families in the last century and of contemporary Turkey has always been the nuclear family (Duben, 1985; Geber, 1989).

As in many Middle Eastern countries, the family in Turkey is the main social security system for the elderly, sick or disabled and the economic refuge for children and youth, the unemployed, and other dependents. Parents are responsible for children until they marry, graduate from school, or become economically independent. Children reciprocate by assuming responsibility for the care of their parents as they grow old (Omran and Roudi, 1993).

A large majority of the people in Turkey has been living in nuclear families, and the extended family has never become a predominant family pattern in actual practice. However, although nuclear families are more prevalent in Turkey (especially in urban and more developed areas and among the better educated), most of the nuclear families can be regarded as socially extended families. Parents and other older relatives may continue to exert influence on the young family's decision-making, especially those living nearby.

Regardless of whether they live in an extended or a nuclear family household, older adults have historically been taken care of by family members (Aytaç, 1995). The elderly are highly respected and valued in Turkish culture. Traditionally, families care for those who can no longer work. However, government is increasingly expected to provide services, such as health care, that are heavily used by the elderly population. In a previous study, Aytaç (1995) found that the incidence of co-residence with elderly people is low (around one-fifth) and the incidence of living nearby is quite high (around one-half), suggesting that Turkish families prefer to have privacy while maintaining close family ties.

Most Turkish families are headed by men (90 percent). The few families headed by women are usually the result of widowhood, divorce, or long-term absence of the male family members. The share of the families headed by widows is increasing slightly because of two factors: life expectancy is increasing faster for women than for men; and women tend to marry men older than themselves (Koç, 1997).

2. Demographic Data Sources

After the foundation of the Republic in 1923, the first population census was held in 1927, and the second in 1935. Till the year 1990, general population censuses have been carried out every five years. Then, it was decided that the censuses be held every ten years. The 14th census of Turkey was held in 2000. In 1997, however, a general population count was held under a special law. The aim was to adjust the number of parliament members and local council members representing the rapidly changing districts before the new elections.

Turkey has a relatively long history of conducting nationwide demographic surveys. The first nationwide demographic survey was held in 1963 and then it quickly became a habit to hold nationwide demographic surveys quinquennially with changing themes. The 1968 and 1973 surveys coincided with the KAP (Knowledge, Attitude and Practice) surveys fielded in many developing countries for the first time. The 1978 survey was part of the WFS (World Fertility Survey) and Turkey was chosen as the pilot country; the 1983 and 1988 surveys evolved within the scope of CPS (Contraceptive Prevalence Surveys); and the 1993 and 1998 surveys were both part of the different phases of the DHS (Demographic and Health Surveys) project. While the dominant themes of the first surveys in the 1960s and the early 1970s were the family and social life, starting from 1978 till today, the dominant themes of the surveys changed towards more family planning and reproductive-health related subjects.

The rest of this paper reviews the existing literature using the results of these demographic surveys of household composition and family structure, including similar new findings from these surveys wherever necessary. Throughout the following sections, the term family structure is used interchangeably with the term household composition. As stated by Koç (1999), the words household and family are typically used interchangeably in the literature and sometimes combined into the term family household. Demographic surveys in Turkey define a household as a dwelling where a person or a group of people usually live and eat together, regardless of whether they are related or not. Therefore, the data from the demographic surveys allow one only to undertake a compositional approach, but nevertheless to take account of the kinship composition of the household according to the head of the household in determining the family structure.

3. Household Composition in Turkey

Table 4 shows that nearly three out of every five households in Turkey were living in nuclear-family households (husband, wife, and children, if any) as early as in 1968. This proportion, however, did not change significantly till the late 1980s; and it has been almost stable again around 67 percent in the last two decades. During this period, extended family households (a married couple living with some relatives of the husband's and/or the wife's) followed rather a steady decline while the dissolved family households (households without a married couple) increased.

The decline observed in the households over time could be attributable to many different factors such as rapid urbanisation, industrialisation, labour-force participation by women, increasing coverage of the social security system, increasing life expectancies, and declining fertility.

Table 4. Percentage distribution of family types, Turkey, 1968–1998.

	1968*	1973*	1978	1983*	1988	1993	1998
Nuclear	59.7	59.0	57.6	61.6	67.4	67.2	67.5
Extended	32.1	32.4	31.9	27.9	22.0	21.6	20.8
Dissolved	8.3	8.6	10.5	10.5	10.6	11.2	11.7

* *Source:* Timur (1972) for 1968, Kunt (1978) for 1973, Ünalan (1988) for 1983.

Table 5. Percentage distribution of detailed family types, Turkey, 1978–1998.

	1978	1983*	1988	1993	1998
Nuclear					
Only husband–wife	8.3	7.9	9.9	11.7	12.6
Husband–wife–1 child	9.5	9.5	12.1	12.8	13.3
Husband–wife–2 children	12.7	14.8	19.1	18.2	18.7
Husband–wife–3+ children	27.1	29.3	26.3	24.6	22.9
All nuclear	57.6	61.6	67.4	67.2	67.5
Non-nuclear					
One-person household	3.0	3.4	3.3	4.4	4.9
One parent and child(ren)	4.8		3.3	4.2	4.7
Dissolved–other	2.8	7.1	4.0	2.5	2.1
Extended	31.9	27.9	22.0	21.6	20.8
All non-nuclear	42.4	38.4	32.6	32.7	32.5
Total	100.0	100.0	100.0	100.0	100.0

* *Source:* Ünalan (1986) and (1988) for 1983.

4. Changes in the Composition of Families

Although the proportion of nuclear-family households has not changed significantly over the last 40 years, a closer look at the composition of these families yields a new perspective (Table 5). The share of families in which couples are living without children increased slightly over the last two decades. With decreasing fertility levels, the number of children nuclear families are having also followed a declining trend and resulted in a lower proportion of nuclear family households with three or more children, and a higher proportion of nuclear family households with one or two children. During the last 20 years, the proportion of couples without children also increased from 8.3 percent to 12.6 percent. During the same period, one-person households also increased from 3.0 percent to 4.9 percent.

The increase in the percentage of husband–wife only families can be attributable to a lot of different factors: 1) the tendency among newly married couples to establish a new household; and 2) to delay their first birth to the later years of fertile life; 3) the increasing availability of housing especially in urban areas; 4) the increase in life expectancy resulting in more elderly couples living alone; and 5) the increasing tendency of grown-up children to leave the household earlier (as the proportion of husband–wife families with children did not change during the last few decades, the effect of this factor could be regarded as very small or it may be compensated for by the tendency to delay marriage).

Changes in the age structure of the Turkish population also contributed to the growing proportion of one and two-person households. However, the cost of living alone, the scarcity of housing, and more importantly, the high value given to the elderly in Turkish culture suppress the potential increase of one-person households.

Table 6. Mean age of the household population by family type, Turkey, 1978–1998.

	1978	1998
Nuclear	23.0	26.7
Only husband–wife	49.6	51.3
Husband–wife–1 child	28.9	29.8
Husband–wife–2 children	23.2	24.3
Husband–wife–3+ children	19.4	21.8
Husband–wife–children	21.3	24.0
Dissolved	26.8	35.4
One person household	64.0	57.8
One parent and child(ren)	20.8	28.7
Dissolved–other	27.3	32.9
Extended	25.8	28.0
Total	24.4	27.6

During the last three decades, average household size decreased from 5.5 in 1968 to 5.2 in 1983 and further to 4.3 in 1998. Three main factors are responsible for this decrease: 1) families have decreased in size due to fall in fertility rates, especially those of higher births; 2) more families started to live in nuclear households; and 3) numbers and proportions of persons living alone have increased.

A useful way to study the linkage between family and age structure is to calculate the mean age of the family members (Table 6). The data show that most people living alone are elderly. However, the decrease in the mean age of the one person households between the two surveys implies that more young people have a tendency to live alone. Overall, the mean age of dissolved family households increased during the period 1978–1998 for all subcategories.

The change in the mean age of the different classifications of the nuclear and extended family households reflected the ageing of the population during the last two decades.

5. Regional Variations

Large variations are observed across geographic regions and between urban and rural areas in Turkey with respect to different demographic and socioeconomic variables. This observation is also true for the variation in family types. According to the 1998 TDHS data, the lowest proportion of nuclear-family households is in the East while the highest proportion is in the South (Table 7). The Southern region was almost consistently found to have the highest proportion of nuclear-family households. On the other hand, the region with the lowest proportion of nuclear-family households changed from the Northern region to the Eastern region. Even though the Western and Southern regions have the highest proportions of nuclear-family households, they differ with regard to the distribution of number of children in these families (Tables 8 and 9); the southern region has higher percentages

Table 7. Percentage of nuclear-family households by region and type of place of residence, 1968–1998.

	1968*	1973*	1978	1983*	1988	1993	1998
West	59.1	61.6	60.9	64.7	73.3	69.6	69.5
South	67.4	66.0	68.5	69.8	71.7	72.3	72.7
Centre	63.6	58.5	55.8	59.8	66.0	65.8	65.4
North	46.3	44.0	50.5	48.7	53.1	54.2	66.2
East	63.0	61.0	50.4	58.3	59.6	65.1	62.2
Urban	64.8	63.2	63.8	67.4	73.5	71.7	72.2
Rural	55.4	55.7	52.0	54.4	58.4	59.1	57.5
Turkey	59.7	59.0	57.6	61.6	67.4	66.8	67.5

* *Source:* Timur (1972) for 1968, Kunt (1978) for 1973, Ünalan (1986) for 1983.

of nuclear-family households with three or more children. Similarly, the nuclear-family households in the Eastern region have a higher number of children than those in the Northern region, especially in the 1998 survey. Figures for the Central region reflect the overall pattern. Nuclear-family households are found to be more common in urban areas. During the last 30 years, the proportion of nuclear-family households in urban areas increased from nearly one-third to three-quarters while those in rural areas did not change significantly.

The trends in different family types over time according to regions are presented in Table 7. The available data and literature allow a comparison of nuclear-family households only over a period of 30 years. Although there are fluctuations in the proportions of nuclear-family households over time in different regions which can be attributable to both real causes as well as to different questionnaires used in these surveys and different methodologies undertaken by previous researchers, the table allows us to draw conclusions about the changing nuclear-family households in different regions of Turkey.

Table 8. Percentage distribution of family types by region and type of place of residence, 1978.

	West	South	Centre	North	East	Urban	Rural	Turkey
Nuclear	60.9	68.5	55.8	50.5	50.4	63.8	52.0	57.6
Only husband–wife	10.1	10.1	9.3	4.6	4.5	9.4	7.3	8.3
Husband–wife–1 child	13.2	11.5	8.6	5.5	5.6	12.1	7.3	9.5
Husband–wife–2 children	16.6	13.7	13.0	8.5	7.0	16.3	9.4	12.7
Husband–wife–3 children	21.0	33.2	24.9	32.0	33.4	26.0	28.0	27.1
Dissolved	11.4	8.2	10.7	9.4	11.2	10.1	10.9	10.5
One person	4.2	2.9	2.6	1.9	2.0	3.4	2.5	3.0
One parent–child(ren)	4.3	3.8	5.2	3.5	6.7	4.2	5.3	4.8
Dissolved–other	2.9	1.5	3.0	3.9	2.5	2.4	3.0	2.8
Extended	27.7	23.3	33.5	40.1	38.4	26.1	37.1	31.9
Total	100.0	100.0	100.0	100.0	100.0	100.0	100.0	100.0

Table 9. Percentage distribution of family types by region and type of place of residence, 1998.

	West	South	Centre	North	East	Urban	Rural	Turkey
Nuclear	69.5	72.7	65.4	66.2	62.2	72.2	57.5	67.5
Only husband–wife	15.0	12.2	15.0	11.7	7.1	12.0	14.1	12.6
Husband–wife–1 child	16.3	14.1	14.7	12.0	6.6	15.0	9.7	13.3
Husband–wife–2 children	22.2	19.2	18.6	20.7	10.8	22.1	11.5	18.7
Husband–wife–3 children	16.0	27.2	17.1	21.8	37.7	23.2	22.2	22.9
Dissolved	13.9	10.8	13.0	10.0	8.5	12.3	10.4	11.7
One person	6.5	3.7	6.2	3.8	2.8	4.9	4.9	4.9
One parent–child(ren)	5.2	5.1	5.1	3.9	3.7	5.2	3.6	4.7
Dissolved–other	2.3	2.1	1.8	2.3	2.0	2.2	1.9	2.1
Extended	16.6	16.5	21.6	23.8	29.3	15.5	32.1	20.8
Total	100.0	100.0	100.0	100.0	100.0	100.0	100.0	100.0

In general, the proportion of nuclear-family households in urban areas appears to increase while there is only a slight increase of nuclear-family households in rural areas. With regard to regions, the biggest increase is for the Northern region followed by the Western. Interestingly enough, even though the biggest change is observed in the Northern region, this region still holds a low proportion of nuclear-family households. The Southern region, which had very high proportions of nuclear-family households even in the 1968 survey, reflects a moderate increase over time and still holds the largest share among all the regions. Despite the decreasing trend of nuclear-family households in the Central region between 1968 and 1978, during the last two decades the proportion of nuclear family households in this region is back at the levels in the 1968 survey.

Tables 10 and 11 present the mean ages of different family types according to geographical regions of Turkey in the 1978 and 1998 surveys. According to the 1998 survey, mean age

Table 10. Mean ages of household members by family types and region and type of place of residence, 1978.

	West	South	Centre	North	East	Urban	Rural	Turkey
Nuclear	25.6	22.8	23.4	20.6	20.0	23.1	22.9	23.0
Only husband–wife	49.6	50.1	48.9	49.2	50.9	46.2	53.5	49.6
Husband–wife–1 child	30.2	26.9	29.1	28.1	26.4	28.0	30.2	28.9
Husband–wife–2 children	23.5	22.3	23.1	23.2	23.4	22.5	24.2	23.2
Husband–wife–3 children	21.2	19.6	19.6	18.3	17.9	19.5	19.3	19.4
Dissolved	32.2	30.4	26.5	23.3	21.4	31.8	24.1	26.8
One person	64.2	69.7	65.3	60.0	56.7	64.8	63.0	64.0
One parent–child(ren)	24.8	22.9	20.9	16.2	18.4	23.8	19.2	20.8
Dissolved–other	30.2	30.6	27.1	26.6	23.1	31.9	25.3	27.3
Extended	29.4	27.2	26.1	24.6	22.5	26.4	25.5	25.8
Total	27.5	24.5	24.8	22.8	21.4	24.7	24.2	24.4

Table 11. Mean ages of household members by family types and region and type of place of residence, 1998.

	West	South	Centre	North	East	Urban	Rural	Turkey
Nuclear	27.9	26.1	28.0	27.8	22.1	26.0	28.4	26.7
Only husband–wife	48.8	50.0	53.8	56.5	54.8	47.4	58.7	51.3
Husband–wife–1 child	29.5	28.5	31.0	31.8	30.0	28.6	33.8	29.8
Husband–wife–2 children	24.6	24.6	23.4	26.1	22.8	23.9	26.1	24.3
Husband–wife–3 children	23.5	22.4	21.9	21.9	19.5	22.0	21.4	21.8
Dissolved	38.5	32.4	36.7	35.5	27.1	34.9	36.5	35.4
One person	58.6	57.6	57.7	56.4	53.7	56.2	61.5	57.8
One parent–child(ren)	30.4	29.2	28.2	31.0	23.3	28.7	28.6	28.7
Dissolved–other	37.4	26.5	36.9	32.1	26.5	33.2	32.3	32.9
Extended	31.0	27.8	28.7	29.3	24.2	28.2	27.9	28.0
Total	29.4	26.9	28.8	28.8	23.3	27.1	28.6	27.6

of nuclear-family households is lower in the less developed Eastern region while there are small differences among other regions. It was interesting to note that, regardless of the family type, mean age of household members living in urban areas was lower than that of rural areas in the 1998 survey. Compared to 1978, rural population had a higher average age in 1998. The mean age of the nuclear-family households in the North increased by more than seven years, while this increase was only two years in the East. With regard to dissolved-family households, the increase in the mean age was especially notable in the Centre and the North. On the other hand, the decrease in the mean age of the household members was largely due to decreases in the West, the South, and the North.

A different approach for examining the relationship between age structure and household structure is presented in Figures 7 to 14. The line graphs show the tendency of household members to live in the respective types of household during different periods of their life.

Figures 7 and 8 present the percentage distribution of the main family types by age group in 1978 and 1998. In 20 years time, there have been important changes with regard to the proportion of people living in specific household compositions in each age group, as a result of the increasing proportion of nuclear-family households and the declining proportion of extended-family households.

According to the 1998 survey (Figure 8), a person is more likely to live in a nuclear-family household in two specific periods of time; while he or she is at school age and during their fertile period. There is a downward trend at the end of school ages and during the early twenties when people are more likely to get married and/or leave the parental home. If the person's age is between 20–24, he or she is also very likely to be living in a non-nuclear-family household compared to other age groups (except the elderly). On the other hand, the elderly population is much more likely to be living in non-nuclear households.

Figures 9 to 11 compare the age specific proportion of the main subcategories of family types between 1978 and 1998. A comparison of the 1998 findings with those from the

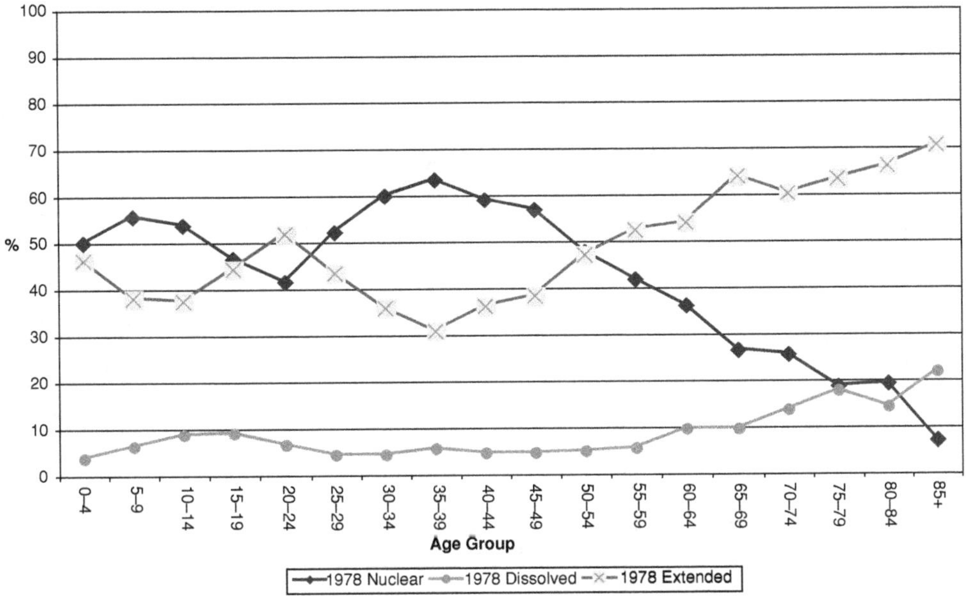

Figure 7. Percentage of household members by family type and age group, 1978.

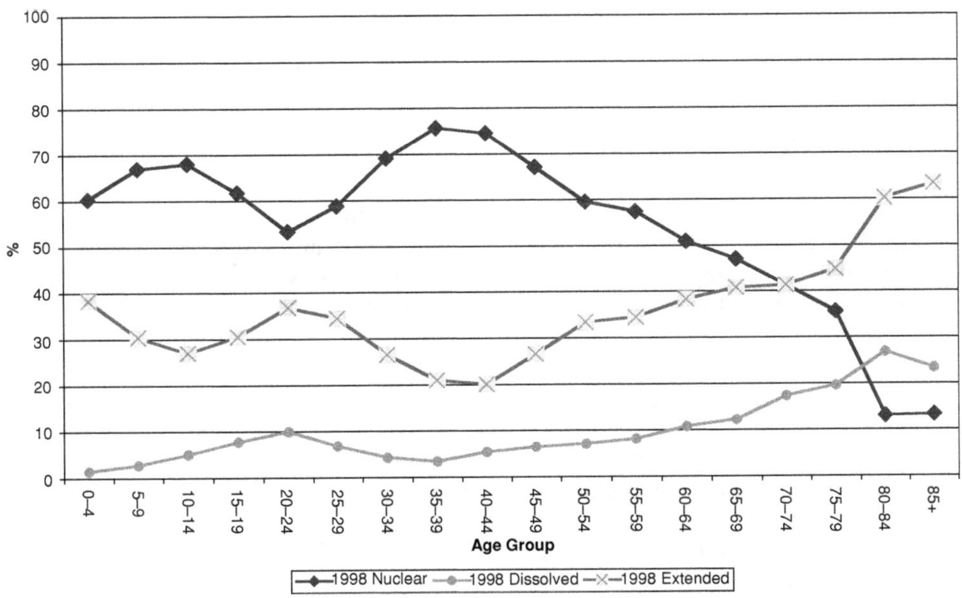

Figure 8. Percentage of household members by family type and age group, 1988.

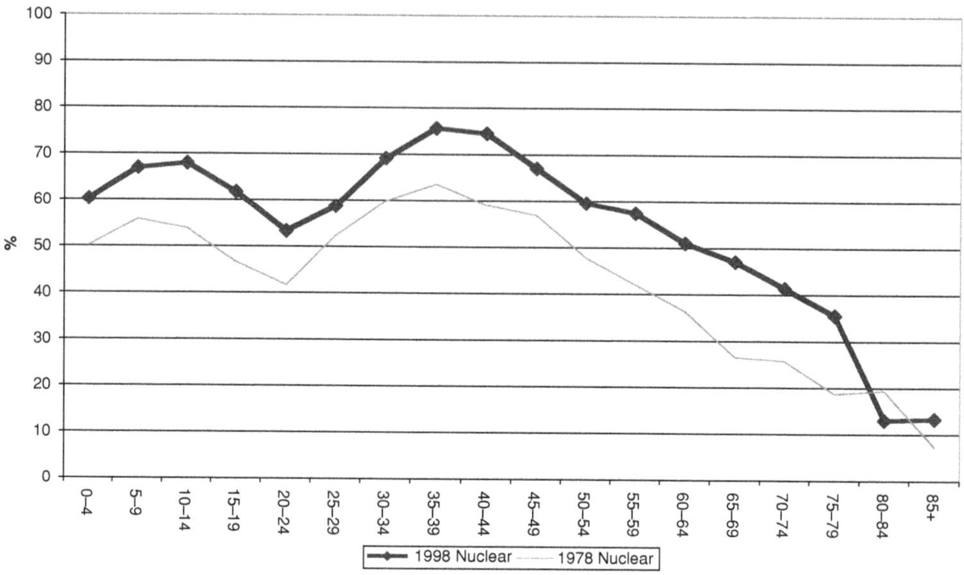

Figure 9. Percentage of household members in nuclear families by age group, 1978 and 1988.

1978 survey (Figure 9) reveals a significant shift of nuclear-family households; regardless of the age group, more people started to live in nuclear-family households. It is interesting to note that the age pattern has largely remained similar, except for the older elderly, while the proportion of members living in nuclear-family households shifted upwards.

For dissolved-family households (Figure 10), the differences were insubstantial but the overall trend by age groups has not changed. The downward trend observed for extended families (Figure 11) was the result of the increasing likelihood of people living in nuclear families. However, the change from extended to nuclear families observed during 1978–1998 affected all age groups equally and the age pattern of extended family households did not change after 20 years (maybe with the exception of older elderly again).

Figure 12 further distinguishes the nuclear families into two groups and the findings from the 1998 survey indicate that people are more likely to be living as couples unaccompanied by children after age 44. Ages between 25 and 44 denote the most likely time of couples enjoying the company of their children. The proportion of couples without children increases rapidly after that age and reaches a peak at age group 70–74. The sharp decline observed afterwards is probably a result of the inevitable death of one partner. Compared to 1978, the proportion of husband–wife-only households in 1998 increased for all age groups between ages 45 to 84.

With regard to dissolved-family households, the age patterns were largely similar except for older ages (Figures 13 and 14); as was pointed out earlier, over time, more elderly

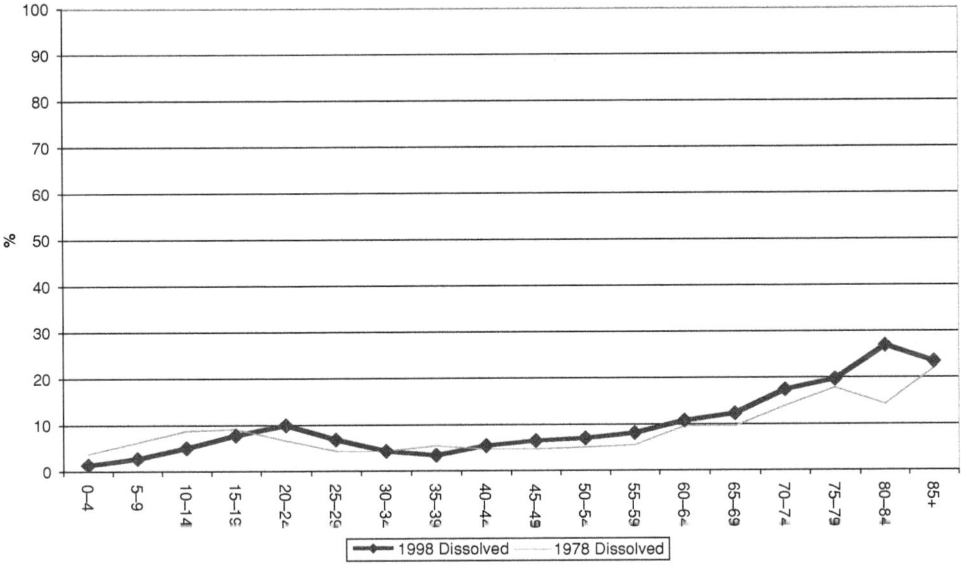

Figure 10. Percentage of household members in dissolved families by age group, 1978 and 1998.

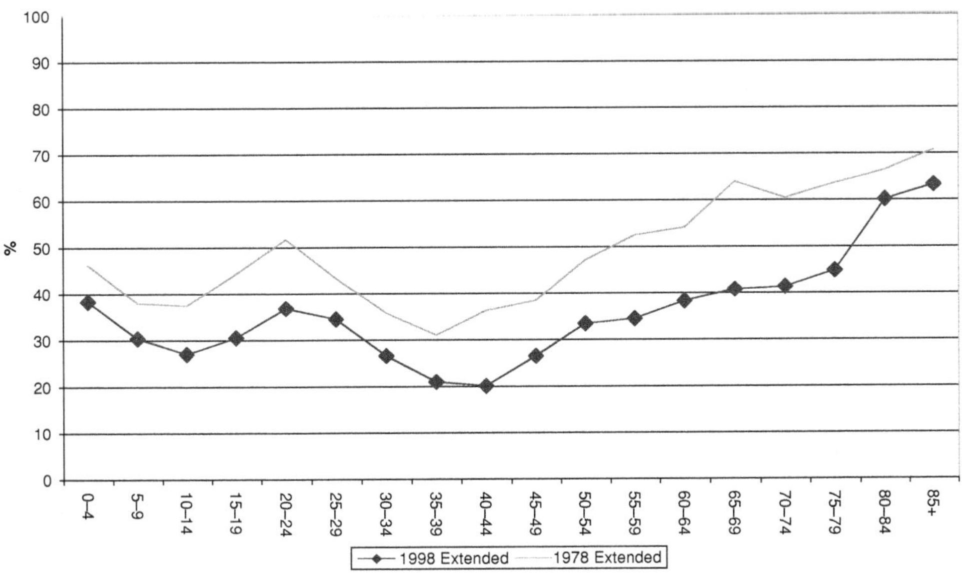

Figure 11. Percentage of household members in extended families by age group, 1978 and 1988.

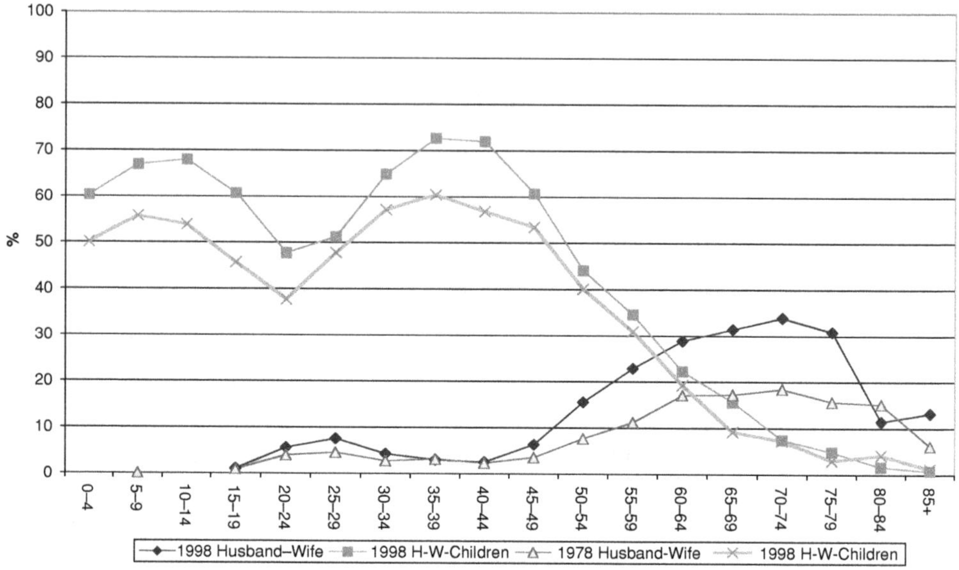

Figure 12. Percentage of household members in husband–wife and husband–wife–children families, 1978 and 1998.

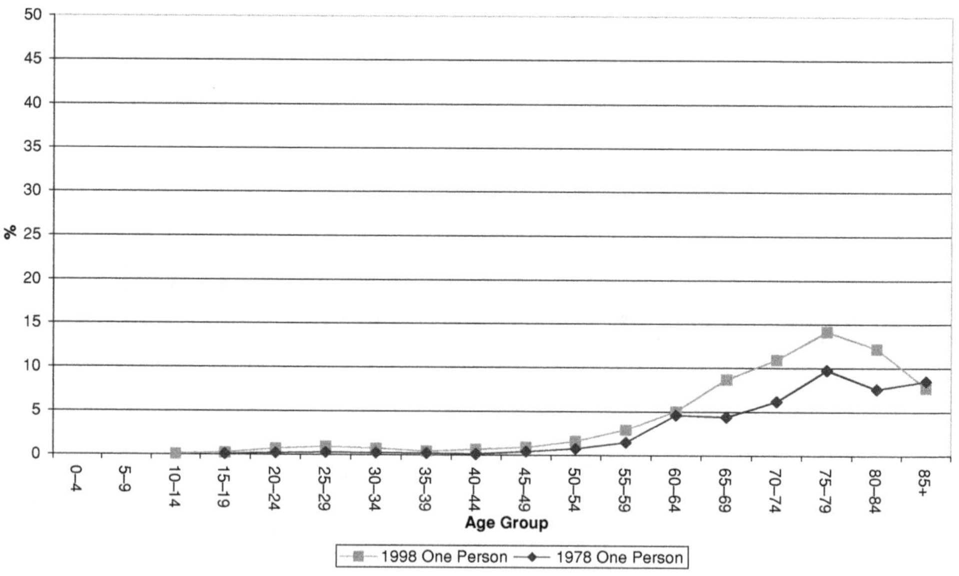

Figure 13. Percentage of household members in one person families by age group, 1978 and 1998.

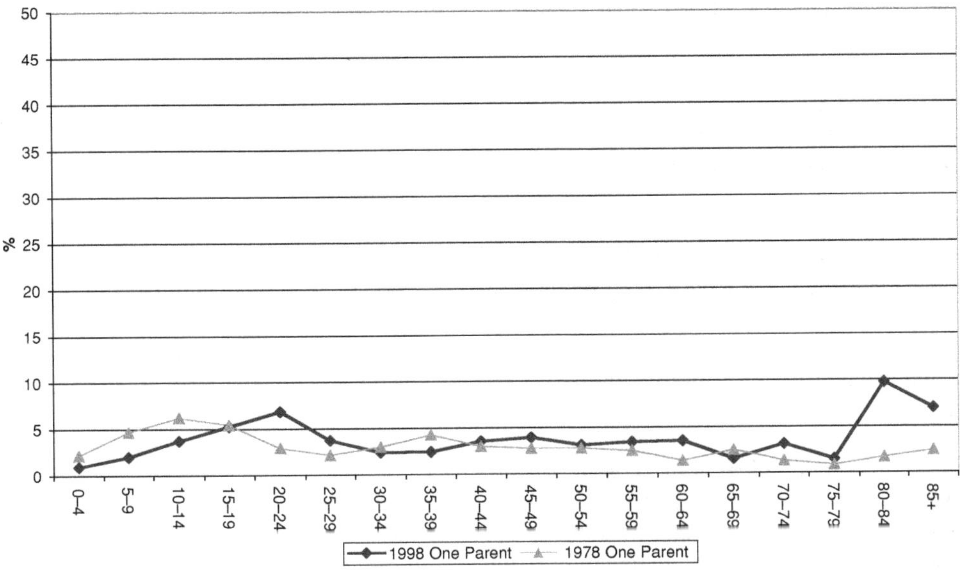

Figure 14. Percentage of household members in one parent families by age group, 1978 and 1998.

people started to live alone and slightly more elderly people were living in one-parent households in 1998.

6. Conclusions

Changes in the size and composition of families have been affected by past demographic changes in Turkish society. Postponement of marriage and childbirth, decreases in fertility and mortality levels all contributed to the changes observed in the size and composition of Turkish families.

The trends in different family types in Turkey were examined in this paper starting from the year 1968 till the last survey of 1998. Using both the existing literature and the available electronic data from the 1978 and 1998 surveys, the trends in the family types were traced over the last three decades. The dominant family type of Turkish society has always been the nuclear family. While three in every five households were classified as nuclear in 1968 survey, it increased to two out of every three households in the 1998 survey. As a result of decreasing fertility levels, the share of husband–wife families within nuclear families slightly increased while that of nuclear families with children decreased. The decrease was especially evident for nuclear families with more children. An important increase was observed for one-person households; and the age composition of these households implied that these households are mostly composed of the elderly and, therefore, contributing to the erosion of traditional extended families. Findings from the 1978 and 1998 surveys implied that there is a decrease in the mean age of one-person families while the other family types have increasing trends.

Analyses of residential differences in family types revealed that there are still diverse characteristics of families across different regions. However, these differences are consistent over time and can be explained with the development level of the particular regions.

An examination of family types by the age structure of the members presented very useful information about in what kind of households Turkish people are living throughout their life. According to the 1998 survey, people are more likely to be living in nuclear households while they are in school ages and between 25 and 44. After age 45, the likelihood of being in a nuclear family decreases quickly and at elderly ages people are much more likely to be living in a non-nuclear family. Comparisons with the 1978 survey also reveals very similar results.

From a methodological point of view, because the existing literature does not have enough material for a detailed comparative analysis and the procedures applied by different researchers to identify the composition of the households may differ, an obvious need is evident to return to the original data files and determine the family types using a standard methodology.

In order to understand the dynamics of different household compositions, more detailed categories are needed, especially for extended families. However, the way the household composition has been identified in the demographic surveys constitutes a serious constraint. Relatives of the household head or the reference person other than children, parents, siblings, bride/groom, grandchildren and in-laws are recorded as other relatives which, in turn, makes it nearly impossible to identify a potential marriage between household members recorded as other relatives of the household head. In order to get a clearer picture of household composition, it is essential to improve the available coding system to allow a detailed identification of kin relation among household members.

References

Aytaç, I.A. (1995), "Intergenerational living arrangements in Turkey." paper presented at the *Annual Meeting of the Population Association of America, San Francisco, CA, April,* Population Research Center and the Center on Aging, University of Chicago, No. 95–9.

Duben, A. (1985), "Turkish families and households in historical perspective." *Journal of Family History*, 10(1): 75–97.

Ergöçmen, B., Hancioglu, A. and Ünalan, T. (1995), *Trends in Fertility, Family Planning, and Childhood Mortality in Turkey: Findings from National Demographic Surveys and Population Censuses*, Ministry of Health Hacettepe University Institute of Population Studies, and Macro International Inc.: Ankara.

Geber, H. (1989), "Anthropology and family history: The Ottoman and Turkish families." *Journal of Family History*, 14(4):409–421.

Koç, I. (1997), "Female headed households in Turkey and socio-demographic and economic characteristics of female household heads." *Turkish Journal of Population Studies*, 19:73–99.

Koç, I. (1999), "Changing family patterns and living arrangements in Turkey (1978–1993)." Paper submitted at *European Population Conference 1999, 30 Aug. to 3 Sept., The Hague.*

Kunt, G. (1978), "Aile Yapisi ve Dogurganlik" (Family Structure and Fertility)." In: *Türkiye'de Nüfus Yapisi ve Nüfus Sorunlari 1973 Arastirmasi, 133–152*, Ankara: Hacettepe University Publications.

Omran, A.R. and Roudi, F. (1993), "The Middle East population puzzle." *Population Bulletin*, 48(1):1–40, (Population Reference Bureau.)

Shorter, F. and Macura, M. (1982), *Trends in Fertility and Mortality in Turkey, 1935–1975, Committee on Population and Demography, Report, No.* 8. National Academy Press; Washington, DC.

SIS (State Institute of Statistics), (1993). 1990 *Census of Population, Social and Economic Characteristics of Population*, Publication No. 1616. State Institute of Statistics: Ankara.

SIS (State Institute of Statistics), (1995), *The Population of Turkey, 1923–1994, Demographic Structure and Development, With Projections to the Mid-21st Century*, Publication No. 1716. State Institute of Statistics: Ankara.

SIS (State Institute of Statistics), (1999), *1997 Population Count: Administrative Division*, Publication No. 2281. State Institute of Statistics: Ankara.

Timur, S. (1972), *Türkiye'de Aile Yapısı (Family Structure in Turkey)*, Publication No. D-15. Hacettepe University Publications: Ankara.

Ünalan, T. (1986), "Some social and economic characteristics of the family types in Turkey with special reference to the Blacksea and Mediterranean regions." *M.A. Thesis*, Hacettepe University Institute of Population Studies: Ankara. (Unpublished.)

Ünalan, T. (1988), "Türkiye'de Çekirdek Aileler." (Nuclear Families in Turkey.) *Turkish Journal of Population Studies*, 10:51–62.

Ünalan, T. (1999), "The status of old age Population in Turkey." Paper presented at the *Conference on the Status of the Older Population: Prelude to the 21st Century, Sion, Dec.*, 13–15, Economic Commission for Europe.

United Nations, (1999), *World Population Prospects, The 1998 Revision, Volume I: Comprehensive Tables*, Department of Economic and Social Affairs, Population Division, United Nations: New York. (ST/ESA/SER.A/177.)

United Nations Population Fund, (1999), *The New Generations, the Family and Society, Population Issues: The Day of 6 Billion*, Online at http://www.unfpa.org/modules/6billion/populationissues/generation.htm

CHAPTER 11. THE 'YOUTH BULGE' AND AGRICULTURE IN THE PHILIPPINES

PETER B. URICH
International Global Change Institute, The University of Waikato, New Zealand

SOCORRO A. GULTIANO
Office of Population Studies, University of San Carlos, Philippines

INTRODUCTION

Population growth in the Philippines remains high and although the transition to lower fertility is occurring it is occurring more slowly than in other countries in the region. There is a 'youth bulge', that is, a preponderance of persons aged between 15 and 30, and this poses a special challenge to the country in terms of the pace and direction of development policy. Little has been written about the implications of the youth bulge on the rural, agricultural sector in the Philippines. This case study explores these issues first at the macro, country level and then links these macro trends with local case study material and in so doing highlights the implications of the macro on the micro and vice versa.

Early results of the 2000 census pegged the population of the Philippines at 76.5 million (NSO 2001). Although the growth rate declined since the 1960s, it still is high (2.36 percent annually in 1995–2000), and it must be noted that the total fertility rates (TFR) in the other countries of the region are lower than that of the Philippines (ESCAP, 1997). The Philippines has lagged behind in fertility reduction. A TFR of 6.0 in 1973 was reduced to 5.1 in 1983, then to 4.1 in 1993, before further declining to 3.5 in 2003 (NSO 2003). The Philippine population remains relatively young: 37 percent of its population is under 15 years of age and another 28 percent is between ages 15–29. Nevertheless, the Philippines is well on its way into the demographic transition and replacement level fertility. However, before this end is reached, the Philippines must first contend with the population's burgeoning number of young adults. "The Philippines is in the midst of a 'youth bulge'", as proclaimed by Xenos and Raymundo (1999: 6). Therefore, it has to deal with the various social and economic implications of this age-structural transition, and needs to deal with them now.

The Philippines recognizes that population dynamics impinge on its goal for sustainable development. One framework for national planning focuses on the integration of population dynamics in agricultural development planning (POPAGRIDEV). This deals with "the determination of the influence of a given population configuration on agricultural

S. Tuljapurkar et al. (eds.), Population, Resources and Development, 203–219.

development, and the influence, effects, impacts of agricultural development on population configuration" (Cabrido 1994). POPAGRIDEV examines population processes (fertility, mortality and migration), and outcomes (size, age–sex structure and spatial distribution) as these relate to various aspects of economic and social development, namely, savings and investment, land, labour and capital utilization, consumption and expenditure, and trade. Specific to the agricultural sector, the concern has always been on the effects of population pressure – owing to high fertility and rapid population growth in and adjacent to the rural areas – on land availability and utilization. Conceptually and empirically, more is known about the causes and consequences of fertility than the causes and consequences of age-structure transformations and geographic redistribution. Consequently, the literature is found wanting insofar as illuminating discussions on age–sex structural changes (particularly in reference to young adults) and development. Little is known about the influence of the youth on agriculture. As a noted rural sociologist observed, the Filipino youth is not as 'visible' in the rural areas as in the urban setting; and hardly enough thought is given to the fact that the youngster in the city may have once been "a barrio lad, or at least his parents were" (Castillo 1979: 159).

It is, therefore, the intention of this chapter to explore the more prominent issues relating to youth and development, with a focus on agricultural development. To carry out the objectives of this study, the paper is organized into two sections. The first section includes: 1) a description of the Filipino youth of today; 2) an examination of recent patterns and trends of agricultural development in the country; and 3) a speculation on the possible associations between patterns of youth behaviour and agricultural development. The second section elaborates on these associations by exploring in case study form: 1) youth mobility in a rural community context; 2) the role remittances might have in altering agricultural development; and 3) how access to the means of production (land) affects the youth in rural areas. Policy insights, where obtainable, are presented in the conclusion. Census data using national summaries illustrate patterns and trends supplemented by other existing literature, while a case study provides insights into the relationships of the patterns and the trends relating to youth and agriculture.

THE FILIPINO YOUTH

On the basis of the Republic Act 8044 (Youth in Nation-Building Act), the National Youth Commission (NYC) of the Philippines defines youth as the population aged 15–29. This age group comprises nearly half of the country's workforce and, therefore, plays a crucial role in "nation-building as catalysts of growth and as the backbone of all development efforts" (National Youth Commission, 1997:1).

In the last three decades the number of youth has gone from 9.7 million in 1970 to 13.7 million in 1980, 17.4 million in 1990, and 21.2 million in 2000. Medium series projections estimate that this number will rise to 26.4 million in 2010, and 29.2 million in 2020 (NSO 1999). As a percentage of the total population, the youth increased from 26.4 percent in 1970 to 28.5 percent in 1980 and 28.7 percent in 1990. Thereafter, this percentage is expected to gradually decline (27.6 percent in 2000 to 26.2 percent in 2020). These figures

Table 1. Population by major age group, Philippines 1970–2020.

Age group	1970[a]	1980[a]	1990[a]	2000[a]	2010[b]	2020[b]
IN THOUS.						
0–14	16,757	20,221	23,994	28,314	28,580	27,661
15–29	9,691	13,698	17,354	21,158	25,724	28,099
30–59	8,560	11,637	16,023	22,467	30,489	38,994
60+	1,646	2,542	3,188	4,565	7,075	10,753
All ages	36,684[c]	48098	60,559	76,504	91,868	105,507
IN PERCENT						
0–14	45.7	42.0	39.6	37.0	31.1	26.2
15–29	26.4	28.5	28.7	27.6	28.0	26.6
30–59	23.4	24.2	26.4	29.4	33.2	37.0
60+	4.5	5.3	5.3	6.0	7.7	10.2

[a]Census of Population and Housing, 1970, 1980, 1990, 2000.
[b]National Statistics Office, 1995 Census-based National, Regional and Provincial Population Projections (medium series), Vol II, Table 2, p. 32.
Includes about 30,000 individuals whose ages are unknown.

illustrate that, indeed, the Philippines is in the midst of a 'youth bulge' (NSO 1999) (Table 1).

Geographic Distribution and Sex Ratios

A larger concentration of youth is in the urban rather than rural areas. In 1990, 30.2 percent of the urban population were persons aged 15–30; the corresponding proportion in the rural areas was 27.2 percent. Whereas the total urban population of the Philippines in 1990 constituted 48.6 percent, the proportion urban in the age range 15–29 was 51.3 percent (NSO 1999).

Females exceeded males among the youth in urban areas (sex ratios of 86.7 in 1980 and 92.5 in 1990). The opposite is true in the rural areas (102.4 and 107, respectively). There is a propensity among young females, to migrate to urban areas. Where frontier-ward migration was dominated by young and middle-aged men (Cruz et al. 1988), urban-ward migration is dominated by young women (Eviota and Smith 1984; Flieger et al. 1976; Gonzales and Pernia 1983; Herrin 1980). Women move to the cities in search of educational and employment opportunities (Abad and Carino 1981).

Literacy and Education

In 1990, literacy rates in the country were already high for both men (93.7 percent) and women (93.4 percent). Among the youth, these rates were even higher: 96.3 and 96.7 percent, respectively. The youth have an educational advantage; by 1990, 26.0 percent of the youth, compared to 16.8 percent of the older population, had been to college.

Table 2. Percent of Population with College Education, by Age and Sex, Philippines 1980–1990.

Age/sex	1980	1990
WITH SOME/COMPLETE COLLEGE		
Total population 20–29	22.6	26.0
Total population 30+	13.5	16.8
Males 20–29	20.7	23.3
Males 30+	14.3	17.2
Females 20–29	24.4	28.6
Females 30+	12.8	16.3
WITH ACADEMIC DEGREE		
Total population 20–29	7.7	11.4
Total population 30+	7.9	9.8
Males 20–29	5.9	8.3
Males 30+	7.4	9.1
Females 20–29	9.3	13.7
Females 30+	8.4	10.7

Source: 1980 and 1990 Census of Population and Housing, National Statistics Office.

More striking is the gender differential in education. Women in the Philippines attain higher levels of education than men (NSO 1994). According to the 1990 census, 40.5 percent of men and 44.9 percent of women aged 20 and older had a college education; 17.4 percent of the men and 24.4 percent of the women obtained a college degree. Over time this gap has widened (Table 2).

Xenos and Raymundo (1999) demonstrated that, from 1960 to 1990, levels of school enrolment among the youth increased significantly. While only 27 percent of males and 25 percent of females aged 15–19 were enrolled in 1960, by 1990 these enrolment rates rose to 40 and 37 percent, respectively. It was estimated that, of those aged 20–24, enrolment rates (generally at the tertiary level) rose from about 11–12 percent for both sexes in 1960 to 22 percent for males and 32 percent for females in 1990. To get a college education, the youth in the towns and villages must move to the city. The corollary to this is, therefore, that more young women migrate to the cities than men.

Employment

It is generally expected that when a person has left school, he or she joins the labor force. An exception are individuals, mostly women, who choose to do housework in their own homes. Recent statistics (1990–1998) from the Income and Employment Division of the National Statistics Office reveal relatively stable labour force participation rates for the youth population aged 15–24 (with rates of 50.0 percent in both 1990 and 1998). However, if disaggregated, an almost perceptible increase in the rates is observed for males (from 60.4 percent in 1990 to 62.0 percent in 1998) and for urban dwellers (44.6 to 47.3 percent), while corresponding decreases are reported for females (38.8 to 37.1 percent) and for rural residents (53.9 and 52.8 percent) (NSO 1999).

In 1990, 44.5 percent of all employed persons held agricultural jobs; this proportion fell to 43.7 percent in 1995 and 39.5 percent in 1998 (NSO 1999). It would appear, therefore, that in terms of labor absorption the agricultural sector of the country had contracted in the 1990s. The youth constituted 32.6 percent of all agricultural workers at that time. The percentage of youth workers was higher among male agricultural workers (32.8 percent) than among females (30.8 percent). The overall participation rate of the youth (ages 15–29) in all types of gainful occupations was 42 percent. Apparently, the participation of the youth in agricultural occupations was much lower than their participation in the other types of occupation. This is to be expected given the proclivity of the youth to reside in urban areas.

Delayed Marriage

There has been a clear trend towards a later age at marriage in the period 1960–1990 (Xenos and Raymundo 1999). For two of the relevant age groups, 15–19 and 20–24, the percent single increased over time. For the age group 20–24, this increase was substantial: from 66 to 73 percent among the males, and from 44 to 56 among females. If similar percentages are computed for the age group 25–29 (but for the years 1970 and 1990), this trend is upheld: the percentage single among males rose from 30.1 to 38.1; among females the percentage single increased from 21.5 to 27.3.

In 1990, the average age at marriage of the rural population (20.2 years), while increasing, remained lower than that of the urban population (21.5 years). Postponement of marriage can trigger concomitant changes in life style. Among the youth of today, living away from the parental home to study or to work is quite common. Cohabitation prior to marriage and premarital pregnancies are no longer rare occurrences. These new norms and life styles can have consequences on the households in the villages where many of the youth come from.

Such a profile of the Filipino youth evidently has implications on the agricultural sector of the economy. These implications are discussed in later sections of the chapter.

PHILIPPINE AGRICULTURE

In the Philippines the influence of population dynamics on agriculture, and vice versa, has received considerable attention. Proof of this is the formulation of a comprehensive framework mapping out population and agricultural development interactions (POPAGRIDEV). It is necessary to take an inventory, albeit partial at best, of the current agricultural situation in the country. The description that follows is derived solely from published data of the 1991 Philippine Census of Agriculture, with accompanying selected information from the earlier agricultural censuses of 1960, 1971 and 1980.

Number of Farms and Farm Area

Consequent to rapid population growth, the number of farms and total farm area in the country has increased over time. In 1960, the reported number of farms was 2.17 million. This number increased to 2.35 million in 1971, then to 3.43 million in 1980. By 1991, there were 4.61 million farms in the country. Correspondingly, total farm area had increased: from 7.77 million hectares in 1960 to 8.45 million hectares in 1971; and from 9.73 million hectares in 1980 to 9.97 million hectares in 1991. During these three decades, the number of farm holdings in the country had increased by 113 percent, but farm area increased by only 28 percent. The average farm area of farming households decreased over time. Although it initially remained relatively stable between 1960 and 1971 (at 3.6 hectares), average farm size shrunk to 2.8 hectares in 1980 then to 2.2 hectares in 1991.

Land Fragmentation

Rapid population growth has contributed to fragmentation of land holdings in the country. The proportion of farms of less than one hectare has increased alarmingly from 1960 to 1991. Where only about one-tenth of the farms were of this size in 1960, already one-third were in this category in 1991.

Land fragmentation is reflected not only in reduced farm size but also in the number of parcels that farming households operate. Between 1960 and 1980, the proportion of farms consisting of two to five parcels declined from 42.4 to 36.0 percent while the proportion with single parcels increased from 55.3 to 62.8 percent. Between 1980 and 1991, however, the trend had reversed: farms with one parcel decreased to 39.3 percent, while those with two to five parcels increased to 59.4 percent. Fragmentation of farm holdings in the 1960s and 1970s had meant giving up some parcels, in many cases as inheritance to older children. In the 1980s, farming of additional parcels, under varying types of tenure, became a farmer's way of coping with the drastic reduction of the size of farm parcels he originally owned or operated.

Land Tenure

Census data on tenurial status, while somewhat difficult to interpret, appears to lend support to the coping mechanism mentioned above. The Comprehensive Agrarian Reform Program (CARP) that was enacted into law in 1988 offers only a weak explanation. Generally, the public lands controlled by central government have been allocated more rapidly than 'alienable and disposable' (privately owned) lowlands. Fifty percent of targeted public lands were dispersed by 1995 versus 35 percent for privately held lands, which in the case of the latter were attained largely from the small-holder sector (Lim 1995). More telling is

Table 3. Agricultural components of population density: 1960–1991

Year	Population[a] / Cultivated Ha.[b]	Cultivated Ha.[b] / Farm Ha.[c]	Farm Ha.[c] / Total Ha.[d]	Population[a] / Total Ha.[d]
1960	4.854	0.718	0.253	0.881
1971	5.710	0.756	0.276	1.193
1980	6.124	0.808	0.316	1.564
1991	6.386	0.953	0.324	1.974

[a]As reported by the National Statistics Office (NSO), 1999 Philippine Yearbook, Table 5.1.
[b]Land planted to temporary crops and permanent crops; 1991 Census of Agriculture: Philippines, Table A.
[c]Total area of all farms reported; 1991 Census of Agriculture: Philippines, Table A.
[d]Total land area of the Philippines (30,747,769 hectares), updated as of 15 December 1996, National Mapping and Resource Information Authority (NAMRIA).

the failure of CARP to redistribute lands above six hectares in the private land sector where only 7508 or 0.8 percent of 845,012 hectares identified for redistribution had by 1995 been disbursed. Meanwhile the allocation of public lands has moved relatively quickly in spite of the fact that many of these lands have important geographical and ecological features that can constrain their full and sustainable development for agriculture, namely they are steeply sloping with fragile soils, and are situated at the headwaters of important river systems.

Intensification and Extensification of Land Use

An inevitable response of agriculture to increasing population pressure is the more extensive and intensive use of land for cultivation. The figures on farm area, average farm size, and number of hectares and parcels per farm discussed above are intuitively illustrative of the ongoing process of agricultural 'intensification' and 'extensification' (Jackson 1992).

The procedure used by Xenos (1998: 51) decomposes population density into different components as follows:

population = (population/cultivated hectares) × (cultivated hectares/farm hectares) × (farm hectares/total hectares)

The ratio of population to cultivated hectares represents 'physiological density'. The ratio of cultivated hectares to farm hectares represents the 'intensity' of use of existing farm lands, while the ratio of farm hectares to total hectares represents the 'extensiveness' of the use of available land in the country (Table 3).

Land used for agricultural purposes has risen from 25.3 percent in 1960 to 32.4 percent in 1991 (growth rate of 28 percent). The proportion of cultivable land in actual cultivation has increased, and even more markedly so: from 71.8 percent to 95.3 percent (growth rate of 33 percent). These rates suggest that agricultural expansion in the country from 1960 to 1991 has been dominated by more intensive use of existing farm lands rather than by the opening up of new agricultural lands. Because 'intensification' seems to have nearly reached its limits – barring the use of better technology – one can probably expect 'extensification' of agriculture to proceed at a faster pace in the future, unless this too is close to its limits (Jackson 1992: 10).

YOUTH IN AGRICULTURE: A CASE STUDY

The youth bulge and the dynamic relationship youth have with education and the labor force can have a profound effect on rural, agrarian communities across the archipelago. The first is definitional, in that some general understanding needs to be established over what constitutes 'youth and agriculture'. Secondly, owing to the better-understood mobility of the youth, particularly toward urban areas, we address the wider impacts this has on local agrarian societies, that is, on those left back in the villages, namely: the youth who choose not to migrate, the more elderly, and the often forgotten children (under 15 years).

Types of Youth in Agriculture

We differentiate between two broad categories of youth (aged 15–30) in rural, agrarian communities. The first consists of youth who, for whatever reason, remain in the rural areas, engage in agricultural work on either their own land, land owned by a parent or land held in long tenancy with the possibility of application for ownership under the laws of agrarian reform. The second group can be broadly characterized as being landless and form a rural agricultural labor pool. The latter group is differentiated from the former by their lack of prospects for attaining land in the future either through inheritance or agrarian reform.

Impacts of Youth Exodus on Agriculture

We have collected data on the nature of the youth exodus from two rural villages: one where wet rice cultivation dominates alongside dry land cropping, and another where dry crop farming in upland marginal lands takes precedence. The villages are both situated in the interior municipality of Batuan of the central Philippine island of Bohol. The two villages flank an extensive area of public and protected land and provide the opportunity to examine where the youth fit in with the exploitation of this resource. Data were collected from each household in these villages in 1992 and through follow-up visits annually thereafter.

In 1992, a total of 187 households were enumerated: 77 in the lowland wet rice cultivated village known as Quezon, and 110 in the dryland cropped village known as Cabacnitan. The *de jure* population totalled 1002 persons: 512 children of living heads of households, 142 other relatives (grandchildren, parents), and 348 who were either the head of household or the spouse of the household head.

Although the land area of the two barangays was nearly identical, their population characteristics were wholly different. For example, the number of households in Cabacnitan was 42 percent greater than in Quezon. Similarly, the population was disproportionately

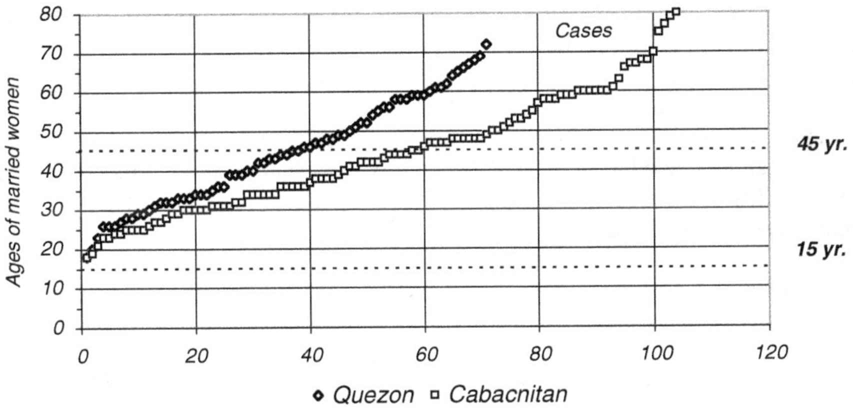

Figure 1. Age and number of all married females by *barangay*.
Source: personal investigation.

distributed. Cabacnitan had 36 percent more people, or 464 full-time residents, in contrast to Quezon's 342.

A difference in number and ages of fecund females was also evident (Figure 1). Cabacnitan had 248 children of between 0 to 15 years of age that represented 3.8 children per fecund female. Quezon had only 143 children, or 3.1 per fecund female. Therefore, not only was the pool of fecund females larger in Cabacnitan than in Quezon, but their fertility rate was higher.

Both villages had about 16 percent of their *de jure* population living outside the village; nearly 95 percent were children (in the youth category) of couples residing in the villages. The other 5 percent consisted of household heads or their spouses who were working outside the villages. Manila was the most common destination for migrants (78 percent of the total, or 99 out of 125). Mindanao was the second most preferred (27 out of 125). Cebu ranked third, with only six migrants originating from the villages.

In terms of migrants' activities, a strong gender differentiation was apparent. Women dominated in areas of higher education (enrolled in courses toward a higher degree) and/or as house helpers. Men were also house helpers, but more commonly worked as drivers or factory workers. A lesser proportion enrolled in higher education. Migrants' work did not result in high levels of remittance to families. Nearly one half of all families reported income through remittances and in some cases this was the sole family income (Urich and Edgecombe 1999).

Youth in Village Agriculture

Thirty households across the age spectrum in both Quezon and Cabacnitan were interviewed, and the lands under their control were surveyed, and their quality assessed. Maps were sketched of each household's fields and questions were asked about land use intensity and methods of cultivation. Although each household was distinctive in all facets of life, we differentiate between three broad types of households. Each type is distinguished in relation to the amount of land households had access to, their tenure situation, the quality of their land, and the household's stage in the life cycle. Households, therefore, had either: 1) relatively large owned holdings, with the family in the later stages of their life cycle; 2) secure tenancies with moderate holdings and primary, elementary or college-age children and hence were in the middle stages of their life cycle; or 3) very little land available to them in any form (owned, tenanted, informally claimed) and were in various stages of the life cycle, but with an emphasis on young families at the beginning of their life cycle.

Family Access to Land

In the studied villages, many clans – excepting those that have claimed land in the public forest – own some land in the cadastrally surveyed areas of the villages, or in those claimed (but as yet not formally titled) after a local forest fire in 1960. These lands were allocated to individual households that constituted the clan. Within each clan there were usually a few families that held tax declarations or titles to the land. Therefore, in most cases, the majority of the families in a clan technically owned no land (did not hold a title to the land in their name) (Veloso 1986).

A clan was comprised by family members, related by blood, who lived in separate and independent households. The type of land that each clan owned was important, for example, irrigated lowland versus xeric uplands. Of additional importance was the number of households in a clan, and the total number of individuals who needed to gain sustenance from that land.

An analysis of land, in relation to family needs, was conducted. Two broad variables were used: land and population, each split into specific categories. Land was divided into its various uses: wet field, dry cropped, and *cogon/talahib* grass. Population was quantified using adult caloric-need equivalents, with each person in a household assigned a value relative to the person's sex and age. A household's land base (and its potential productive capacity) was broadly compared with the number of calories that the household members required for basic sustenance.

One extended family, the Bolon's, is used as an example for the way that land resources were distributed in relation to family size. As a clan, the Bolon's own a total of 30.24 hectares of land, half of which is *cogon/talahib* grass-covered uplands. The remaining lowlands were cultivated to nearly one hectare of unirrigated rice, slightly more than eight hectares of corn, and just over five hectares of coconuts. The Bolon's were the largest clan in the study area with a total of 92 individuals living in 18 separate households. Disregarding the unproductive *cogon/talahib* grass uplands, the Bolon clan relied on less than 1600 square meters (0.16 hectares) per individual for sustenance, or a mean of 1.68 hectares per family. The Bolon clan's cumulative adult equivalent was 58.2. However, the clan's land base was not evenly distributed among all its households (Table 4).

For example, one of the households headed by a young, newly wedded couple (Bolon 5) had the least land at their disposal, while the clan's patriarch (Bolon 1) owned over three

Table 4. Lands owned, tenanted and occupied by various households of the Bolon clan, in hectares.

Total	Owned	Tenant	Occupied
Bolon 1	3.37	0.00	3.36
Bolon 2	0.42	0.00	1.10
Bolon 3	0.00	0.0	0.82
Bolon 4	0.00	0.71	0.0
Bolon 5	0.00	0.24	0.0

hectares that consisted of unirrigated rice land (0.30 ha.), corn land (1.88 ha.) and upland (1.20 ha.). Bolon 1 also had over three hectares of dry cropped land in the public forest zone. One of the households, Bolon 3, only had access to land in the public forestland. The trend was for older members of the clan to own land, with the middle-aged and youth tenanting from them. Younger households had a higher proportion of land held as either tenancies – usually of degraded dry cropped lowlands as in the case of Bolon 5 – or as insecure illegally occupied land in the forest zone (Bolon 2 and 3).

Adult caloric-need equivalents were not correlated with the area and type of land available to each family. Bolon 1 had an adult equivalent of only 2.73 (mature-aged couple), while Bolon 3 and Bolon 5 (in the youth cohort) had poor access to good land and relatively high adult equivalents. In general, needs in terms of calories for any one household in the clan *were not* associated with the disposal of the clan's land. Allocation of lands was therefore not equitable. Elderly, dominant families maintained the ownership of land, and allocated it to younger families as they saw fit. Trends were for the clan's younger couples (the youth) to receive less land in lieu of their ability to work as wage laborers. Middle-aged households have better access to land.

Life Cycles and Land Use

The interaction of life cycles and household structure affected land management decisions. For example, households ranged in composition from young couples with one or more children, to families through their reproductive years who supported children in school, and finally, more aged households whose children had completed school. The youth were represented in the first group as heads of household and their spouses, and in the third group as grown up children in aged households. In many cases, the latter enjoyed a shift in wealth flow from children to parents. Within each group of household, access to land of various quality affected land use decisions.

Intensive Land Use

In Cabacnitan, 17 of the 20 households interviewed continuously cropped their uplands. Many were young couples (15–30 years of age) with growing families. They lived close to the forest margin, and often cultivated land in the forest zone. Youth, lack of access to land, and poverty are correlated (Lynagh and Urich 2002). As a proportion of all those cultivating land in the public domain, the youth clearly outnumbered all others. In Quezon, only 2 out of 25 farmers surveyed, who had uplands under their control, cultivated them continuously. The situation for these households in many ways matched conditions of the 17 families of Cabacnitan who cultivated continuously. One case was a refugee family of eight, and a second, was a single, male headed household, with five young children supported by a land base of only 0.31 hectares of tenanted rainfed lowland and a small adjacent upland plot. In cases such as Quezon's, and with older couples in Cabacnitan cultivating in the public forest zone, a common characteristic is not a family headed by someone in the youth cohort but rather one child (or often times more) in the youth cohort still attending school.

Moderately Intensive Land Use

Those who cultivated their uplands more moderately, that is, on more regular rotation, were almost exclusively located in Quezon. Uplands were viewed as a food safety net, rather than as an integral part of the household's economy. Upland plots were cultivated about every 10 years. Moderate use of fragile lands was conducted by families with more adequate access to other lands or with ready access to remittances from migrant members of the household. The youth *per se* were not well represented in this group but they did play a role: families in this group tended to be slightly older with youth to support in college. The cost of financing college influenced land use decisions.

Extensive Land Use

As a percentage, the *barangay* of Quezon had more older couples whose children had been educated and were now living or working outside the province and could in many cases remit. When a household owned a considerable area of land relative to others, distinctive land management decisions had been made. Cultivation of fertile plots was sometimes discontinued because the aggravation and production costs outweighed the benefits or perceived needs of the household. Thus, only a minimum amount of land was cultivated, and this was usually the land nearest the house.

In some cases, lands were cultivated solely to assist the village's underprivileged. Members of the land-owning households seldom participated in these 'make work' projects. Importantly, the landowner felt obliged to create an opportunity for the less fortunate youth to work and harvest a crop of rice to feed their young families. No matter how benevolent the owners felt, however, they still claimed their typical five-sixths of the harvest and left the rest for the labourers.

Commonly, those with excess land changed to less labor-intensive cropping systems such as fruit and other trees. Areas planted to coconut trees increased among the older, more secure households. Degraded lowland corn land was planted with coconuts. Economic trees planted in the uplands reflected a 'comfort zone' for the households in terms of the area of their land holdings and their perceived need to cultivate uplands during drought. This

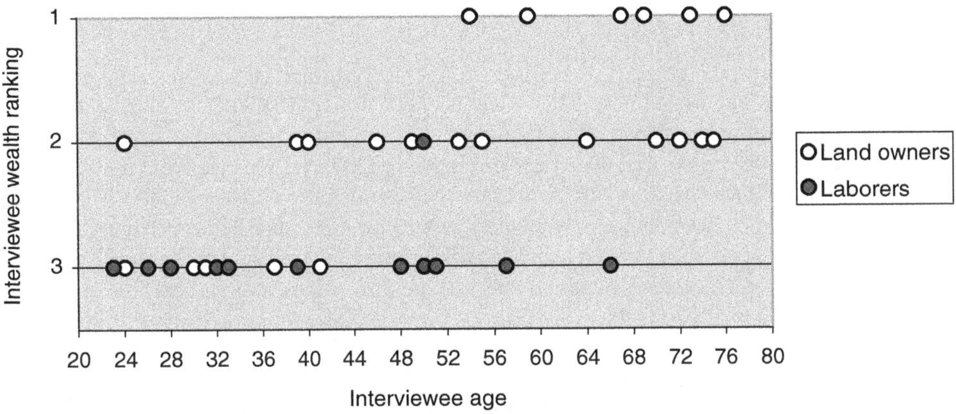

1. The wealthiest segment of the population
2. The 'middle' class
3. The poorest group

Figure 2. Outcome of wealth ranking exercise with families in village of Cabacnitan associated with a local irrigation system. Survey included land owners and laborers in the system.

latter point becomes more critical with time, as the land base available for the cultivation of small grains and root crops for daily consumption and opportunities to labor in their production continually diminish. This relates to the next point we wish to make regarding the so-called 'landless youth'.

Landless Youth

There are members of the youth population who do not own or tenant land in the 'alienable and disposable' portion of the villages. Their prospects for inheriting land, securing long-term tenancy or a piece of land through agrarian reform is virtually nonexistent. This segment of the youth population is often poor and works as a laboring class in the villages (Figure 2). Wages, however, are not sufficient to maintain their families so they resort to cultivation of public lands for sustenance.

Lynagh and Urich (2002) have found in their research on the laborer population in the village of Cabacnitan that, as a cohort, the youth were unlikely to change their habits in terms of cultivating public lands regardless of changes in economic conditions in the village. The public land is perceived as a 'free' resource and one that should be exploited. Conservation is viewed as secondary to family survival. This is evident from the fact that a majority of the affluent farmers (of the older age groups) who also cultivate land in the public domain would abandon their claims if requested by the government.

Summary

Types of land use and its intensity were indicative of any number of socioeconomic factors that affected individual households. There was a correlation between the amount and type of land owned, and the type and intensity of land use. Implementation of more sustainable land use strategies, i.e., tree planting, was more common in Quezon. This was due in part to the security of tenure Quezon's land owners had to their land. Large areas of Cabacnitan were owned or cultivated by households who placed tenants on individual hills and many of these tenants were young and with limited access to other lands. These tenancies were frequently rotated (as the youth migrated between the village and urban areas), and this militated against the planting of any long-term economic crops such as coconuts, fruit or timber trees.

The life cycle of a household influenced land use and intensity. Families that owned and tenanted land in Quezon were generally older and had educated their children and were now receiving remittances. This permitted the land manager freedom to plant economic trees in lands they would have otherwise cropped to cover school expenses. Cabacnitan's families were considerably younger, and they needed to cultivate land intensively.

Families compete for land, either as tenancies, or as new parcels in the forest zone and even in degraded cadastrally surveyed land. This reflected not only the limited extent of available alienable and disposable land but also a decline in agricultural productivity and removal of cultivable land from daily, subsistence crop production. It tends to be the young farmers aged between 15 and 30 with young families who make use of poor, marginal land or illegally occupy public land. They may be in line to inherit land from elders but with fragmentation and involution the prospects of supporting a family on inherited land are waning rapidly.

In terms of the youth exodus and the youth bulge, these have a very critical impact on land use intensity and relate also to the family life cycle. Families exploit their land more when they have children to support in school or university in order to shoulder these additional expenses. Depending on the extent of their landholdings, they then withdraw land from production as children remain in the city and get employment. Wealth flow then reverses as children remit to their parents in the villages. Remittances can influence how intensely land is cultivated, as does the cost of labor. A way around the latter is to over-exploit the very young and the old.

YOUTH AND AGRICULTURE

The youth is a relatively mobile and highly educated segment of the population with a proclivity to migrate to urban areas. Rural–urban migration, however, is highly selective of young women who go to the cities to pursue a college education and/or to find employment in the growing service sector. The consequence of this selectivity is that disproportionately more young men, who either do not have the opportunity or inclination for further studies, or have difficulty finding a job in the city, remain in or return to the rural areas. Searching for viable means of livelihood, these young men pursue agriculture. However, agricultural land is growing scarce. The result, therefore, will be more intensive use of existing farmlands and encroachment into forestlands, natural reserves, and other marginally productive lands. Such a situation will likely perpetuate and increase poverty and inequity in the rural areas. Moreover, indiscriminate use of land undermines the goal of sustainable development.

A positive contribution of the youth bulge to agriculture may come in the form of increasing remittances of migrants who have found work in the cities to their families in the rural areas. The pervasiveness of this practice, the volume of money and goods remitted, and the manner in which these remittances are put to use will determine the extent of usefulness of this practice and the contribution of migrant youth to rural development.

CONCLUSION

The youth bulge in the Philippine context has implications for rural as well as the urban sectors of society. While issues of an increasingly urbanized youth have been privileged in the literature, this should in no way discount the impacts in rural areas. This is especially important in a country like the Philippines which still has a large, politically powerful rural peasantry.

The complexity of the rural situation mirrors that of the more urbanized youth. The rural youth are also mediating the boundaries between a more traditional rural existence and the forces of modernization. Youth engaged in agriculture are generally poor and this can be the result of several converging issues.

First, they may not have had access to high school and college education which then places them at a severe disadvantage when choosing a life course. If they are restricted in their urban-ward migration for this reason, they remain in rural areas where they are once again hindered by a lack of access to land and other work opportunities.

Of critical importance in the access of land is internal clan relationships and resource allocation. Significant also is the problem of land conversion to non-grain crops by the more elderly and relatively land-rich families. As more land is converted to pasture and tree crop production or is withdrawn from cultivation (abandoned), the land base available for the youth to work on, and the community to gain basic food needs, declines.

Remittances are also an increasingly important factor for those in the rural community that have educated their children and are now experiencing a reversal in wealth flow. The application of these funds to further development of the family widens the gap between those that have education and have migrated and sent remittances, and the youth who have remained local and are trying to engage in agriculture.

As far as policy is concerned, it is important for government to take heed of the fact that youth in the rural areas consist of two groups: those who leave the area and those who stay. Reasons for out-migration should be addressed. Education is among these reasons. It is a common perception that the public high school curriculum in the country places heavy emphasis on academic preparation for tertiary education. While this is not altogether futile, perhaps more effort should be placed in making education, even at the secondary level, more relevant to people's living conditions and work prospects. In regard to work conditions of migrant youths in the cities, care should be taken that fair and equitable treatment is given to these people. Incomes earned by these migrants find their way to needy families in agrarian communities.

Policy should also address the needs of youth who stay in the agricultural sector, that is, the youth with relatively easy access to land and those without. The latter group is particularly vulnerable to ongoing communist and insurgency movements in the rural areas. Therefore, sustained and intensified efforts should be placed in the implementation of the Comprehensive Agrarian Reform Program. At the same time, nature conservation measures must be effectively put in place to prevent further degradation of the environment. To raise family incomes, especially among the landless youth, opportunities for off-farm employment must be generated.

This study has recognized the importance of family and clan-based relations in the allocation and use of land. For agricultural programs to be effective, more studies will have to be undertaken on the existing inheritance system and kinship structure in various parts of the country as these affect agricultural opportunities for the youth.

References

Abad, R. and Carino, B. (1981), *Micro-Level Determinants of Migration Intentions in the Ilocos: A Preliminary Analysis*, pp 2–5. Institute of Philippine Culture, Ateneo de Manila University.

Cabrido, C. Jr. (1994), *Integration of Population Dimension in Agricultural Development Planning*, IPDP Project, National Economic and Development Authority.

Castillo, G. (1979), *Beyond Manila: Philippine Rural Problems in Perspective*, p 159. IDRC: Ottawa, ON.

Cruz, M.I., Zosa-Feranil, I. and Goce, C. (1988), "Population pressure and migration: implications for upland development in the Philippines," *Journal of Philippine Development*, 15(1):15–46.

ESCAP. (1997), *Population Data Sheet*, United Nations: Bangkok.

Eviota, E. and Smith, P. (1984), "The Migration of Women in the Philippines." In: *Women in the Cities of Asia: Migration and Urban Adaptation*, J. Fawcett, S. Khoo and P. Smith, eds. pp 165–190. Westview Press: Boulder, CO.

Flieger, W., Koppin, B. and Lim, C. (1976), *Geographical Patterns of Internal Migration in the Philippines: 1960–1970*, National Census and Statistics Office.

Gonzales, V. and Pernia, E. (1983), "Patterns and determinants of interregional migration." In: *The Spatial and Urban Dimensions of Development in the Philippines*, E. Pernia, et al., eds. pp 303–317. Philippine Institute for Development Studies.

Herrin, A. (1980), *Population and Development Research in the Philippines: A Survey*, Philippine Institute of Development Studies.

Jackson, R.T. (1992), "At the wrong end of the cul-de-sac? Philippine landuse since 1903." *4th International Conference on Philippine Studies, Canberra.*

Lim, E. (1995), *CARP: Crunch Time! Intersect*, 9(6):4–5 and 18–19.

Lynagh, F. and Urich, P.B. (2002), "A critical review of buffer zone theory and practice: a Philippines case study." *Society and Natural Resources*, 15:129–145.

National Statistics Office. (1970), *Census of the Philippines: 1970*, Bureau of the Census and Statistics. Bureau of Printing: Manila.

National Statistics Office. (1980), *Census of the Philippines: 1980*, Bureau of the Census and Statistics. Bureau of Printing: Manila.

National Statistics Office. (1990), *Census of the Philippines: 1990*, Bureau of the Census and Statistics. Bureau of Printing: Manila.

National Statistics Office. (1994), *National Demographic Survey 1993*, Tables 2.4.1 and 2.4.2.

National Statistics Office. (1999), *1995 Census-based National, Regional and Provincial Population Projections*, Vol. 2.

National Statistics Office. (2003), *Philippine National Demographic and Health Survey 2003: Summary Report, Manila*, p 4.

National Statistics Office. (1999), *Philippine Yearbook*, pp 650–651. Tables 15.2 and 15.2a.

National Statistics Office. (2000), *2000 Census of Population and Housing Highlights of the Preliminary Results, Manila.*

National Statistics Office. (2001), *2000 Census of Population and Housing, Manila.*

National Youth Commission, (1997), "Overview of the major factors affecting the youth." Manuscript prepared for the *Multisectoral Consultative Summit on the Medium-Term Youth Development Plan, Clark Field Pampanga, Feb. 26–28.*

Urich, P. and Edgecombe, M. (1999), "Bohol's indigenous social institutions: a development perspective." *Philippine Quarterly of Culture and Society*, 27:178–201.

Veloso, G. P. (1986), "Agricultural lessor-lessee relationships in the province of Bohol: implications to the teaching of land reform: a proposed instructional program." *Ph.D Thesis*, University of Bohol.

Xenos, P. (1998), "The Ilocos coast since 1800: population pressure, the Ilocano diaspora, and multiphasic response." In: *Population and History: The Demographic Origins of the Modern Philippines*, D. Doeppers and P. Xenos, eds. p 51. Ateneo de Manila University Press: Manila.

Xenos, P. and Raymundo, C.M. (1999), "The modern profile of Filipino youth" in *Adolescent Sexuality in the Philippines*, C.M. Raymundo, R.P. Xenos and L. Domingo, eds. UP Office of the Vice Chancellor for Research and Development: Quezon City.

CHAPTER 12. SINGAPORE'S CHANGING AGE STRUCTURE: ISSUES AND POLICY IMPLICATIONS FOR THE FAMILY AND STATE

ANGELIQUE CHAN
National University of Singapore

Introduction

This paper discusses the implications of population ageing for policies regarding support for older persons in Singapore. Singapore is one of the fastest ageing populations in Asia as a result of extremely successful family planning policies instituted in the 1960s (Saw 1999). As such, Singapore is an excellent case study of the effects of rapid ageing on State policies. Currently 7 percent of Singapore's population is over the age of 65; however, by 2030 this will increase to 19 percent (Inter-Ministerial Committee on Ageing Report 1999). As Pool discusses in Chapter 2 of this volume, age-structural transitions affect supply and demand factors within a society, differentially putting pressure on key life-cycle stages and consequently policies catering to the needs of populations at specific ages. In this paper I discuss the policy implications of population ageing in the areas of financial security, employment, living arrangements, and health care. I combine this discussion with illustrations using recently available longitudinal data for Singapore. My focus is on the changing needs of the elderly over time and the need for policy makers to take these changes into account when developing policies for older adults. Given increases in life expectancy, the importance of changing needs within the span of 'old age' (from age 65 onwards) becomes even more pertinent. Longitudinal data provide for more accurate projections of the changing needs of the elderly over time (Andrews and Hermalin 2000). For example, housing and health care policies for the aged may have to be fine-tuned so as to cater to the specific needs of the young-old (65–74) versus the oldest-old (75+).

The family has traditionally been the main source of support for the elderly in Asia, and Singapore is no exception. In recent decades, academic debate has centered upon gauging the effect of modernization or industrialization on levels of familial support (Cowgill and Holmes 1972; Martin and Kinsella 1994). This interest intersects with the awareness of policy-makers in Asia that traditional family support of the elderly may decline in future. There is some evidence of declining levels of familial support in countries such as China and India due to the effects of massive rural–urban migration of young adults, and changes in the occupational structures (World Bank 1994). This has led some policy makers to decry the influence of 'Westernization' and its by-product, individualism. However, there is also evidence to suggest that the Asian family is adapting to changing economies and that elderly well-being is not declining. Research on intergenerational transfers in Indonesia,

S. Tuljapurkar et al. (eds.), Population, Resources and Development, 221–242.

Malaysia, the Philippines, Singapore, Taiwan, and Thailand, has shown high-levels of intergenerational support for elderly parents either via co-residence or the transfer of goods and services (Knodel 1997a, 1997b; Hermalin 1997, Chan 1999a, 1999b; Ofstedal, Knodel and Chayovan 1999; Chang 1999).

The goal of most Asian governments is to develop programs that do not undermine family support which is perceived as far superior to state-based support, both morally and financially (Ofstedal, Chayovan, Chan et al. 2001). In Singapore, the government has strongly upheld the belief that familial support of the elderly is the ideal. To a certain extent, we see that there remain very high levels of family support, as measured by co-residence rates. Approximately 85 percent of Singaporean elderly live with at least one child. However, familial support cannot be measured by co-residence levels alone (Hermalin 1997). Indeed, one of the main reasons for persistently high co-residence rates in Singapore may be high housing costs, in addition to high levels of filial piety. To look at family support per se, we need an understanding of the intergenerational transfers that occur within the family in order to better judge whether the elderly are being taken care of.

The Singapore family is under increasing pressure to support its older members given the low fertility rates since 1975. As will be elaborated upon later, Singapore has been experiencing below replacement fertility since 1975. Although current elderly cohorts have large numbers of children, low fertility rates since 1975 will translate into fewer children available to support older parents in future.

The rapidly changing age structures of many Asian nations require that we understand the complex changes that occur within families as economies develop. The challenge for policy makers in all of these countries is to develop sound systems to ensure economic and social well-being of the elderly. This means developing a delicate mix to enhance levels of familial support while at the same time ensuring individual preparation for old age. Understanding the role of policies in ensuring old-age support necessitates the recognition that there exists a politics of ageing which involves the allocation of resources by age (Rappa 1999). Singapore is not a welfare state and individual responsibility is the mantra underlying most, if not all, policy-making decisions. This attitude is reflected in the types of policies implemented, for example, the Central Provident Fund described later in this paper.

In the following section, I review the changes that have occurred in Singapore's age structure since independence in 1965. I then discuss various policies for financial security, employment, living arrangements, and health care, and highlight the main areas of concern. Where possible, I provide illustrations of the changing characteristics of elderly Singaporeans over time, using data from a longitudinal survey carried out between 1995 and 1999.

Singapore's Changing Age Structure

At present, Singapore's population stands at 4.02 million, with an annual growth rate of 1.8 percent for resident (citizens and permanent residents) Singaporeans (Singapore Census of Population 2000). The growth rate of non-residents was much higher, 9 percent (see

Table 1. Singapore's population and growth rate, 1871–2030.

				Growth rate (%)		
Year	**Total pop^n**	**Residents**	**residents**	**Non- pop^n**	**Total Residents**	**Non- residents**
1871	97,111	–	–	–	–	–
1881	137,755	–	–	3.6	–	–
1891	181,612	–	–	2.8	–	–
1901	227,592	–	–	2.3	–	–
1911	303,321	–	–	2.9	–	–
1921	418,358	–	–	3.3	–	–
1931	557,745	–	–	2.9	–	–
1947	938,144	–	–	3.3	–	–
1957	1,445,929	–	–	4.4	–	–
1970	2,074,507	–	–	2.8	–	–
1980	2,413,945	2,282,125	131,820	1.5	–	–
1990[1]	3,047,132	2,735,868	311,264	2.4	1.8	9.0
2000	4,017,733	3,263,209	754,524	2.8	1.8	9.3

[1] Includes resident population residing overseas.
Source: Census of Population 2000.

Table 1). This "non-resident" group consists mainly of migrant labourers ensconced in lower-level occupations, such as construction or foreign domestic workers.

Older Singaporeans, aged 65 and above make up 7 percent of the population, and by the year 2030 they are expected to make up 19 percent of the population. This translates into an increase from 235,000 elderly in 1999 to 796,000 elderly in 2030 (IMC 1999). The oldest-old population (85 and above) is experiencing the fastest growth at 6.1 percent per year. As a result of improvements in sanitation, medical technology, and public health awareness, life expectancy has risen in Singapore to 76 for men and 79 for women (IMC 1999).

As shown in Table 2, this ageing of the population has had the cumulative effect of raising the median age of the population from 19 in 1957 to 34 in 2000 (Singapore Census of the Population 2000).

The ageing of Singapore's population was primarily motivated by a fertility decline beginning in the late 1950s (Fawcett and Khoo 1980; Saw 1975). Total fertility rates declined rapidly in Singapore from an average of 6.5 births per woman in 1957 to 1.9 births per woman, or below replacement fertility, in 1975. Since 1975, fertility rates have remained below replacement level despite attempts by the government to increase fertility. Incentives to increase fertility have had only modest success. The total fertility rate remains below replacement level (1.8 in 2000).

The remarkable speed by which fertility declined has been primarily attributed to an aggressive government campaign to reduce fertility, which began in the late 1960s.

Table 2. Median age for Singapore (1911–2000).

Census year	Median age
1911	28
1921	28
1931	26
1947	23
1957	19
1970	20
1980	24
1990	29
2000	34
2010	37
2020	39
2030	41

Source: Singapore Census of Population 2000 and the Inter-Ministerial Committee Report on the Ageing Population (1999).

(Leete (1994) disputes this arguing that fertility declines were just as rapid for overseas Chinese in Malaysia in the absence of aggressive family planning policies.) The Singapore Family Planning and Population Board was established in 1966. Its main goal was to make family services physically and financially accessible to all couples in the reproductive ages and to reduce fertility and the rate of population growth (Fawcett and Khoo 1980). A number of disincentives to having children were created. These disincentives were intensified during periods when fertility was appearing to increase. For example, following an increase in the crude birth rate between 1970 and 1972, disincentives to having more than two children were intensified in 1973. These disincentives included increases in government hospital delivery fees for higher birth order children; abolition of paid maternity leave after two children; no priority for Primary One registration for the third and subsequent child unless one parent underwent sterilization; no income tax relief for fourth and higher birth order children (Saw 1975). Due to the high premium placed on education, these disincentives led many women to opt for tubal ligations and abortions, while some simply refused to bear more children. In 1973 there was a 54 percent increase in the number of sterilizations, and a 38 percent increase in the number of abortions performed in the previous year (Saw 1975). Singapore reached replacement level fertility in 1975, and fertility rates soon fell below replacement level in 1977. The total fertility rate has remained below replacement level and stood at 1.6 births per woman in 2000 (http://www.singstat.gov.sg).

The realization that the Singapore population was not reproducing itself led the Government to reverse its population planning policy. Since 1987, Singaporeans have been encouraged to have three or more children if family finances permit (Lee, Alvarez and Palen 1991). Numerous financial incentives and other incentives including priority schooling, have been provided to couples having three or more children. (Lee, Alverez and Palen (1991) provide an excellent review of the types of incentives provided to increase fertility from 1987 to

1991.) Recently, the government has instituted the provision of a 'Baby Bonus' for parents having a second or third child. Under the new Baby Bonus scheme, a Children Development Account (CDA) will be opened for a family once a couple has a second or third child. For the second child, the government will contribute $500 per year into the account and up to another $1000 each year to match contributions from the parents – dollar for dollar.[1] For the third child, the government will double its contribution to $1000 per year, and up to $2000 in matching contributions annually. This baby bonus will stop when the child turns six as the government states that it is the primary responsibility of parents to provide for children. The money in the CDA can be used to pay for the development and education of all children in a family.

In addition to the baby bonus, mothers will get eight weeks of paid maternity leave for their third child, instead of just for the first two children as provided for in Singapore's Employment Act. The government will pay the wage cost of maternity leave for the third child (to be capped at S$20,000) so as to ease employer burden. In addition to these financial incentives, the government aims to create a total environment conducive to raising a family such as making child-care centres more easily available and affordable, and creating family-friendly work arrangements in the civil service.

The policy implications of this dramatic drop in the fertility rate range from supply and demand issues affecting youth (see Jones 2000 in this collection for a discussion of the interrelationship between declining fertility and education policies in Singapore) to supply and demand issues affecting the elderly. Thus, while the government is actively promoting an increase in fertility rates, the concern with policies for the elderly is also receiving focused attention.

There are several economic and social policy implications of a rapid ageing population. One of the most apparent economic implications is the increase in the old age dependency ratio that will occur within the next 30 years. In the year 2000, one elderly person was supported by 9.8 working persons. This will decrease to 3.5 working persons per elderly person, in the year 2030 (Vasoo, Ngiam and Cheung 2000). Therefore policy makers are concerned with the need to ensure financial security of elderly persons either through retirement savings or an extension of employment years. Social implications include the need for appropriate housing and health care for current and future cohorts of elderly. In this report, the following six areas were highlighted as foci for policy development for elderly: employment and employability, housing and land use policies, financial security, social integration of the elderly, healthcare, and cohesion and conflict in an ageing society. The various areas of concern are neatly spelled out in an Inter-Ministerial report on ageing that was published in 1999. (For a review of government policies for the aged see Vasoo, Ngiam and Cheung (2000).)

The government's review of the population planning policies began in 1984 with the establishment of the Inter-Ministerial Population Committee (Vasoo, Ngiam, and Cheung

[1] Unless otherwise specified, dollar amounts reported are in Singapore dollars. At the time of writing, US$1 equaled S$1.8 (26 January 2001).

2000). At the same time, the Government began analyzing the impending shift in Singapore's age structure from a youthful population to a mature and aged population. As early as 1982, a high-level committee on 'Problems of the Aged' was formed to address the long-term impact of population ageing. Since then various policy recommendations have been implemented such as changing the provident fund contribution rates for older persons, legislation on minimum standards for old age homes, increasing elderly dependants' tax relief, and legislation on filial piety, i.e., the Parental Maintenance Act (Vasoo, Ngiam and Cheung 2000). In June 1988, the National Advisory Council on the Aged (NACA) was formed to address ageing issues in Singapore. Various policies have been implemented as a result of the NACA's recommendations, including an increase in the retirement age from 60 to 62 on January 1st 1999. Most recently in 1998, the Inter-Ministerial Committee on Ageing Population was established to examine six main areas related to the elderly: employment and employability, housing and land use policies, financial security, social integration of the elderly, healthcare and cohesion and conflict in an ageing society.

The rest of this paper will center upon a discussion of financial security, employment, living arrangements, and health care policies for the aged. I highlight the key issues in these areas and provide illustrations using the data described below.

Data

The recent availability of longitudinal survey data has facilitated in-depth analyses of the complex processes that occur within families as economies develop. I use data from the 1999 survey of Transitions in Health, Wealth, and Welfare of Elderly Singaporeans. The data are nationally representative and form a source of panel data. In 1995, 4750 individuals of age 55 and above were interviewed concerning a variety of issues including demographics, work, intergenerational support, income, health, and voluntary activities. In 1999, researchers attempted to re-interview as many of the original respondents as possible. This project was a collaborative effort between researchers at the National University of Singapore, the Ministry of Community Development and Sports (Singapore), and the Population Studies Center of the University of Michigan (USA). Funding for the project was provided by the National University of Singapore. Taking into account the mortality rate for this age group (4 percent per year) and other losses to follow-up (including moves and severe health impairments impeding interview), we managed to re-contact 42 percent of the original respondents. This resulted in a total sample size of 1981.

The re-interview collected information on changes that occurred to the individual during the four-year period. In addition, the project team supplemented available information by including a detailed questionnaire module on Central Provident Fund usage and holdings. The Income and Assets module was also expanded. The re-interview collected in-depth information on health and health care services usage. We also expanded our database on available kin networks.

The data allow for analyses of transitions in health, wealth, and welfare of elderly Singaporeans between 1995 and 1999. As such, investigators can now make more definite claims with regard to the direction of causality in numerous processes, e.g., the relationship between health and retirement. Previous studies that used cross-sectional data could not separate cause from effect, e.g., does poor health cause retirement, or does retirement cause health status to deteriorate? By collecting information on when a person retires and subsequent information on health, researchers can test the hypothesis that retirement results in a deterioration of health status. Indeed Andrews and Hermalin (2000) note that in order to develop sound policy, there is a need for longitudinal data and given the subtlety of the information needed, this usually points to a panel design with re-interviews rather than reliance on retrospective reporting of complex histories.

Financial Security

In economic terms, below replacement fertility levels since 1977, translate into a decrease in the number of working age persons (15–64) per older person (65+). As Jones (2000) notes, however, the dependency ratio is a crude indicator. Increased productivity of the working age population could translate into a decline in 'real' dependency, even as 'demographic' dependency rises. Old-age economic support is an issue that the Singapore government has been concerned with since independence. Singapore's Central Provident Fund (CPF) was instituted in 1955 as a mechanism to provide Singaporeans with financial security in old age. Since its inception, the CPF has evolved into a savings system that "changes the whole concept of social security from provision for retirement to provision for life" (Choon and Low 1996). It remains a mandatory savings scheme for all employees and employers in Singapore. Some observers have questioned, however, whether individual CPF savings will be sufficient for support in old age. As Asher (1996) notes, government estimates show that by 2003 the net balance of members' accounts, after withdrawal for housing, investments, and other schemes, will not reach the minimum balance. Thus many elderly will have to resort to private savings and family support (Shantakumar 1999). Singapore citizens and permanent residents are eligible to be covered under the CPF scheme. For employees (both public and private) participation is mandatory. Contribution rates vary depending on the state of the economy. Current rates are shown in Table 3. Since 1994, self-employed individuals can elect to open a CPF account.

A CPF member has three savings accounts into which their contributions are transferred automatically; an Ordinary account, a Medisave account, and a Special account. Ordinary account savings can be used to purchase property, approved investments, insurance, to pay for children's tertiary education, and to top-up spouse's or elderly parents' CPF accounts. Medisave account savings are for meeting hospitalization and medical expenses and to buy medical insurance. As of 1 July 2001, CPF members must keep a minimum of $26,000 in their Medisave account to pay for medical expenses.

Savings in the Special account can only be withdrawn at age 55. This age used to coincide with the retirement age. Although the retirement has been gradually raised to 62, the age limit for CPF withdrawals has not been altered. There have, however, been substantial changes in how an individual can make use of his/her CPF savings. Individuals are now able to invest more of their CPF savings in government-approved schemes; however, they

Table 3. CPF rates by type of account as of 1 January 2001.

Employee age (years)	Contribution by employer (% of wage)	Contribution by employee (% of wage)	Total contribution (% of wage)	Credited Into		
				Ordinary account %	Special account %	Medisave account %
35 & below	16	20	36	26	4	6
Above 35–45	16	20	36	23	6	7
Above 45–55	16	20	36	22	6	8
Above 55–60	6	12.5	18.5	10.5	0	8
Above 60–65	3.5	7.5	11	2.5	0	8.5

Source: Central Provident Fund Board (http://www.cpf.gov.sg).

must maintain a minimum sum for old age financial needs. As of 1 July 2001, individuals must maintain a minimum sum of $70,000 in his/her Retirement account to be used for old age income. This minimum sum will be increased by $5000 every July until it reaches $80,000 by the year 2003 (Central Provident Fund Board, 6 December 2000). This target figure will yield a monthly annuity of $613 for approximately 20 years from retirement (IMC 1999). Alternatively, the sum can be used to buy a life annuity with an insurance company or deposited with a bank which will yield a monthly income. The government is presently trying to encourage CPF members to purchase annuities which would provide them with a regularly monthly income in old age. The take-up rate, however, has been slow prompting the government to offer several seminars to inform the elderly about annuities.

Recently, the CPF Board has announced a new policy whereby the percentage of savings channelled into the Ordinary, Special, and Medisave accounts each month will be linked directly to age (CPF Board News Release 30 May 2000). The CPF is constantly evolving policies to cater to the needs of the population. Table 3 shows the contribution rates for the various CPF accounts as at 1 January 2001. These rates are adjusted depending on the growth of the Singapore economy. For example, during the 1997 Asian Economic Crisis, contribution rates by employers were decreased by 10 percent points. The CPF is a highly efficient savings mechanism for the Singaporean government. Currently, Singapore's CPF has the highest coverage of any retirement plan in Asia. The CPF contributes between 16.3 percent and 30.4 percent to the gross national savings rate (Asher 1995).

Cohort differences in CPF coverage are significant. A much higher percentage of those aged 55 to 59 years old in 1995 are covered by the CPF (52 percent) compared to those aged 70–79 (25 percent). The percentage of elderly aged 80 and above in 1995 that have CPF accounts is even lower, 14 percent (Chan 1999). Of those with CPF accounts, the majority, 31 percent, had a total of under S$5000 in their account (see Table 4). One-fifth

Table 4. Total CPF savings at age 55 among elderly who were age 59 and above in 1999.

Total CPF savings (in Singapore dollars)	Percent
< 5000	30.7
5,000–9999	7.8
10,000–19,999	12.0
20,000–29,999	5.6
39,000–39,999	4.7
40,000–49,999	3.2
50,000–99,999	6.1
100,000–149,999	2.7
150,000 & above	2.4
None	24.9
Total	100.0

Source: 1999 Transitions in Health, Wealth, & Welfare of elderly Singaporeans:1995–1999.

of the elderly in this sample had no savings left in their CPF account. The Government has also been providing cash top-ups to individuals. Recently, all Singapore citizens at least 21 years of age on 31 December 2000, and who had contributed at least S$100 into their CPF accounts, were eligible to receive top-up payments of between S$500 and S$1700. The actual amount received is dependent upon employment status, monthly salary, and type of housing (Straits Times 3 January 2001).

Given the low coverage rates and savings amounts among the current generation of elderly, this generation is less likely to rely on CPF savings and more likely to rely on family support. In fact, 87 percent of this cohort of elderly (age 59 and above in 1999) are living in households in which they are not the main breadwinners. Most of these elderly report receiving money regularly from children; 79 percent of Chinese elderly, 63 percent of Malay elderly, and 44 percent of Indian elderly (Chan 2001).

While these cohort differences exist, individual changes in income status over time need to be taken into account. Policies regarding the economic well-being of Singapore elderly are based on the idea that family support will supplement individual saving and satisfy the elderly individual's needs. Often, the underlying assumption is that the needs of the elderly either remain static or decrease over time. However, little work has been done on the relationship between needs and actual income changes at older ages in Singapore. Most analyses of elderly economic well-being have used cross-sectional data. Unfortunately these data do not allow for individual-level comparisons of changes in economic well-being over time. As shown in Table 5, there is substantial variation over time in actual income reported by elderly Singaporeans; 20 percent of elderly respondents reported a decrease in income whereas 32 percent reported an increase in income level. A total of 48 percent of respondents reported

Table 5. Changes in income for elderly respondents between 1995 and 1999: Singapore.

Monthly individual income in 1995[a]	Change in income			Total	Income level	
	Decrease	Same	Increase		1995	1999
< $500	–	62.8	37.1	100	52.0	46.9
$500-999	37.3	35.6	27.2	100	30.5	27.6
$1,000-1,499	48.0	21.4	30.6	100	11.0	13.2
$1,500-1,999	57.5	21.3	21.3	100	2.6	5.4
$2,000+	50.7	49.3	–	100	4.0	6.9
Total	20.1	48.4	31.5	100	100.0	100.0

Chi-square = 539.04 (df = 8), $p < 0.001$
Source: Chan, A., Ofstedal, MB, and A. I. Hermalin. Forthcoming. Changes in Subjective and Objective Measures of Economic Well-Being and Their Interrelationship Among the Elderly in Singapore and Taiwan. Social Indicators Research

no change in income over the four-year time period. (I make an implicit assumption that the cost of living did not change over the two survey periods for each country. In periods of relative economic stability, the cost of living is unlikely to change dramatically over a four-year period. However the Singapore survey did not occur during such a period and it is likely that the Asian economic crisis that began in October of 1997 affected the cost of living between the survey waves, possibly increasing costs for some things and reducing costs for others.)

These shifts in actual income levels are correlated with changes in marital status, employment status, and living arrangements of the elderly. For example, retirement and widowhood are significantly associated with decreases in actual income over time. (For a detailed analysis of the determinants of these changes see Chan, Hermalin and Ofstedal forthcoming.) These results suggest the need for specific policies to cater to certain sub-groups of elderly persons, e.g., widowed or retired elderly. As such, blanket policies that refer to the elderly in general will be of limited success unless target populations are identified and their concerns addressed. Attention has also to be paid to the fact that the measurement of income levels and adequacy among the elderly has many facets.

Economic well-being has objective and subjective components. For example, elderly with similar actual income levels may perceive their income to be inadequate due to large financial or familial commitments.

One possible measure of economic 'need' is the perception of income adequacy. Table 6 compares levels of perceived income adequacy reported in 1995 by respondents aged 55 and above at that time, with levels of perceived adequacy reported by the same individuals four years later. Of those respondents reporting enough money left over in 1995, 72 percent reported a worsening of perceived adequacy in 1999. Of those respondents that perceived much difficulty with income adequacy, 91 percent reported an improved status by 1999. These results show a substantial change in levels of perceived income adequacy over time which may reflect changes in levels of need.

Table 6. Changing levels of perceived income adequacy between 1995 and 1999.

Level in 1995	Change in adequacy: Better	Same	Worse	Total	Perceived adequacy: 1995	1999
Enough with money left over	—	28.0	72.0	100.0	10.5	21.0
Just enough, no difficulty	21.0	63.2	15.8	100.0	79.9	61.8
Some difficulty	68.2	26.7	5.2	100.0	7.8	14.7
Much difficulty	90.6	9.4	—	100.0	1.9	2.5
Total	24.0	55.7	20.6	100.0	100.0	100.0

Source: Chan, A., Ofstedal, MB, and A. I. Hermalin. Forthcoming. Changes in Subjective and Objective Measures of Economic Well-Being and Their Interrelationship Among the Elderly in Singapore and Taiwan. Social Indicators Research

The hypothesis that economic needs of the elderly change over time is further supported by the finding that perceived and actual income levels are not highly correlated (see Table 7).

Among elderly who experienced an actual increase in income between 1995 and 1999, only one-third reported an increase in perceived income adequacy. In fact, 15 percent of elderly reported a decrease in perceived income adequacy. For those elderly whose actual income level remained the same, 22 percent reported an increase in perceived adequacy, and 20 percent reported a worsening of perceived income adequacy. Finally, among those elderly who experienced a decrease in actual income over the time period, 17 percent reported an increase in perceived income adequacy, and 54 percent reported that their perceived income adequacy remained stable.

There are several possible explanations for this lack of correlation between perceived adequacy and actual income. A change in living arrangements could mitigate the relationship between actual income levels and perceived income adequacy. For example, the transition into a larger household could result in fewer of the elderly individual's needs being met.

Table 7. Association between change in actual income and change in perceived income adequacy: Singapore, 1995–1999.

Change in income between 1995 and 1999	Change in Perceived Adequacy: Better	Same	Worse	Total
Increase	31.4	53.1	15.2	100.0
Same	21.7	58.7	19.5	100.0
Decrease	16.7	54.2	29.2	100.0
Total	23.8	55.0	20.2	100.0

Chi-square = 46.1 (df = 4), $p < 0.0001$

Source: A., Chan, MB, Ofstedal, and A. I. Hermalin. Forthcoming. Changes in Subjective and Objective Measures of Economic Well-Being and Their Interrelationship Among the Elderly in Singapore and Taiwan. Social Indicators Research

There are several policy implications of these results. Firstly, the perception of income adequacy and changes in actual income are not highly correlated. Policy makers interested in facilitating financial planning for old age need to be sensitive to this disjuncture when interpreting data for policy-planning purposes. Elderly who perceive income adequacy to have increased, may actually have experienced a drop in actual income. For example, elderly who retire and move in with children may experience a decrease in actual income, but receive transfers-in-kind from children to compensate.

Secondly, there is a great deal of instability in actual income levels over time for Singapore elderly. Our findings suggest that actual income levels vary substantially during old age. This is primarily an outcome of one's health needs, work status, education levels, and living arrangements. How each of these factors operates is still unclear but an understanding of the underlying mechanisms will allow policy planners to better predict which individuals are likely to experience declines in actual income. This allows better targeting of policies towards needy populations.

Employment and Employability

Over the past few years, labour and skill shortages have driven government policy to recruit foreign talent. In addition, the Government has begun to emphasize the re-training of older workers. However, many older workers have not sought active re-employment (Shantakumar 1999). This shift towards realizing the potential of older workers is a recent trend. Singapore continues to have a mandatory retirement age that was raised from 60 to 62 on 1 January 1999 with the expectation of raising it to 67 years in the future. At present, employers tend to be hesitant about hiring older workers and this has led to public debates on the issue (Straits Times 16 July 1999). Among the current generation of elderly (aged 59 and above), 16 percent are employed. Employed status includes individuals that are currently holding a job, those that are holding a job but temporarily not working for various reasons, and individuals that are working in family business but are not getting paid. Male elderly are more likely to be employed (28 percent) compared to female elderly (8 percent). The panel data shown in Table 8 reveal the sharp drop in number of elderly individuals employed over the survey period. This is a product of the mandatory retirement age of 60 applicable to this cohort. Recently, the Government has emphasized the importance of a shift in attitude among employers towards hiring older workers (*Straits Times* 16 July 1999). Job advertisements in Singapore newspapers tend to specify preferred ages which disadvantage older applicants.

This runs counter to recent Government initiatives to keep older workers in the work force. As shown in Table 8, the percentage of unemployed older workers rose over the 1995–1999 period for both male and female elderly. This suggests a growing pool of elderly persons who would like to work if given the chance to do so.

Among those elderly who continue to work, reasons for working include financial need, the need to remain active, and to prevent boredom (see Table 9). There are some ethnic differences in reasons given. Chinese elderly are most likely to cite feeling bored as a

Table 8. Employment status of elderly Singaporeans, 1995 and 1999.

Employment Status	Total (%) 55 & above (1995)	Total (%) 59 & above (1999)
Total		
Employed	27.4	16.2
Unemployed	1.5	5.1
Economically inactive	71.1	78.6
Male		
Employed	44.3	27.8
Unemployed	1.8	4.9
Economically inactive	53.9	67.3
Female		
Employed	12.1	7.7
Unemployed	1.3	5.3
Economically inactive	86.6	87.0
Total (N)	4750	1981

Source: Transitions in Health, Wealth, and Welfare of Elderly Singaporeans: 1995–1999.

reason for continuing to work whereas Malay and Indian elderly are most likely to cite a feeling that they can still lead an active life. A quarter of the Chinese elderly sample report needing money as a reason for working, compared to 16 percent of Malays and 22 percent of Indians.

Table 9. Main reason for continuing work after retirement by ethnicity, 1999.

Reasons for working after retirement	Ethnic Group				
	CHINESE	MALAY	INDIAN	OTHER	Total
Need money for own and family expenses	26.6%	15.8%	22.2%	33.3%	25.1%
Need money for future financial security	7.1%	7.9%	11.1%		7.4%
Not enough support from children	6.0%	18.4%	11.1%	33.3%	8.0%
Saving for something specific	.4%				.3%
Feel that can still lead an active life	27.0%	36.8%	33.3%		28.3%
Interested in job	3.6%		5.6%		3.2%
Feel bored	28.6%	18.4%	11.1%	33.3%	26.4%
Others (specify)	.8%	2.6%	5.6%		1.3%
Total	252	38	18	3	311
	100.0%	100.0%	100.0%	100.0%	100.0%

Source: Transitions in Health, Wealth, and Welfare of Elderly Singaporeans: 1995–1999.

Housing and Land Use Policies

Living arrangements of the elderly in Asia are often used as an indicator of elderly well-being by researchers and policy makers. In early studies of living arrangements of the elderly, co-residence levels were used to make indirect evaluations of levels of family support of the elderly in Asia (Martin 1988). Since then, research has focused more on the 'function' of the family as opposed to the 'form' (Hermalin 1997). The understanding is that although levels of co-residence between elderly and adult children may decrease over time, intergenerational transfers across households may ensure that the well-being of the elderly does not decline. Singapore has one of the highest co-residence rates in Asia. In 1995, 86 percent of elderly with at least one adult child, live with at least one adult child (Chan 1997). This high co-residence rate is in part a function of availability of children with whom to live with, and a culture that stresses filial piety. In addition, high housing costs in Singapore make it more economical for adult children to co-reside with older parents.

Government policies have also encouraged co-residence both as a moral obligation of adult children to older parents, and as an attractive financial arrangement. The Singapore government has instituted a variety of tax and financial incentives such as tax reliefs and priority housing to adult children who live with, or nearby, elderly parents. In addition, the government has recently introduced housing options for those elderly that choose to live alone or only with a spouse.

Currently, as shown in Table 10, 3 percent of both male and female elderly (aged 55 and above) live alone. Males are more likely to be currently married, and hence a larger proportion, 69 percent live with a spouse compared to female elderly, 28 percent. Male elderly are more likely to be living with unmarried children (60 percent) whereas female elderly are more likely to be living with married children (50 percent). Female elderly are also more likely to be living with grandchildren (46 percent) compared to male elderly (29 percent). Only 2 percent of the elderly in the sample co-reside with siblings. However, these living arrangements are not static. During the period of old age, elderly can live in a variety of living arrangements.

Table 10. Type of living arrangements for respondents aged 59 and above in 1999.

Household members	Males (n = 2058)	Females (n = 2692)
Living alone	3.3	3.6
Spouse	69.1	27.5
Unmarried children	60.1	48.4
Married children	32.3	50.0
Grandchildren	28.9	46.4
Siblings	2.2	1.5
Total		

Source: Transitions in Health, Wealth, and Welfare of Elderly Singaporeans: 1995–1999.

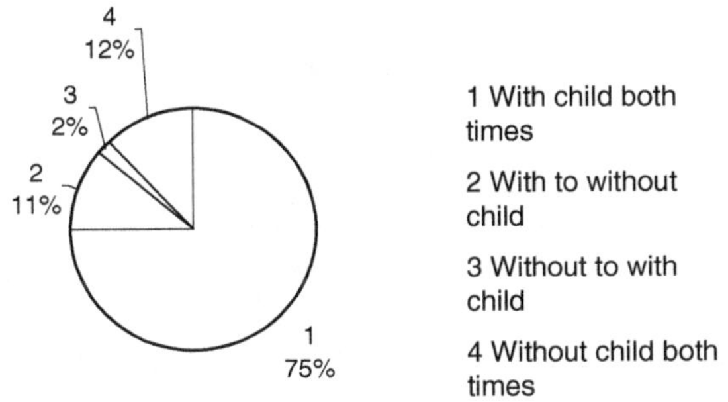

Chart 1. Changes in living arrangements between 1995 and 1999.
Source: A. Chan, E. Frankenberg, and M.B. Ofstedal. 2001. "Stability and Change in Living Arrangements of the Elderly in Southeast Asia." Submitted for Publication.

Chart 1 shows the amount of change that has occurred between 1995 and 1999 for this group of elderly respondents; 75 percent of respondents lived with an adult child at both times, 2 percent transited to living with a child, and 11 percent transited to not living with a child. The later group most probably consists of respondents whose adult children have left home. In a companion paper, we investigate the determinants of these transitions in living arrangements to isolate some of the significant explanatory factors (Chan, Frankenberg and Ofstedal 2001). We find that although co-residence is common, much of it may reflect a situation in which young adult children have not yet left home. As they age and leave school, co-residence becomes less common. In Singapore, the institution of two years of national service for male adults to be completed when they are 18 and 19 years old, later and increasing for children have created a situation where adult children remain dependent on parents for longer periods of time. On average, males do not graduate from university until the age of 24 whereas females graduate at 22 years old (Chan, Frankenberg and Ofstedal 2001).

Current government policies promote 'ageing in place'. This policy underlies the recent efforts to retrofit older housing estates with elder-friendly features such as lifts at every floor, non-slip flooring in bathrooms, corridor railings, and wheelchair access (IMC 1999). However, it remains to be seen whether future cohorts of elderly Singaporeans will opt for privacy and prefer to live alone or with a spouse only. Although modernization theory (Cowgill and Holmes 1972) would predict a decrease in co-residence rates as modernization occurs, Singapore's unique situation may dispel this theory. Consistent and far-reaching government programs encouraging family support, in addition to high housing costs and lower marriage rates, may result in the maintenance of high co-residence levels.

Health Care

One of the recommendations outlined in the recent Report of the Inter-Ministerial Committee on the Ageing Population (1999), was the need for more research on health issues concerning the Singapore population. Health care policies for the elderly have been receiving a lot of attention in the past year. The government has been actively promoting awareness of health care issues among the elderly population in general. At the same time, elderly without health-care insurance have been urged to open medical insurance (Medishield accounts) with aid from the government. In December of 2000, the government announced it would spend S$110 million to pay basic Medishield premiums for two years for Singaporeans aged 61 to 69. Premiums range from S$96 a year for those aged 61 to 65, to S$132 for elderly aged 66 to 70. Elderly who fail to qualify for coverage, because of pre-existing medical conditions, will receive the equivalent of the premium in the form of a Medisave top-up (*Straits Times* 7 January 2001). Approximately 65 percent of eligible elderly enrolled in the program (*Straits Times* 8 January 2001). With the inclusion of this group of elderly, approximately 90 percent of elderly Singaporeans between the ages of 61 and 69 now have health care benefits.

The Ministry of Community Development and Sports (MCDS) has recently begun a five-year master plan for elder care which involves increased funding to Voluntary Work Organizations, funding a series of programs for the elderly, and spending S$15 million on a five-year public education program. Funding will be provided to set up support services for frail elderly. These support services will be run out of three centres with the help of twelve case managers. These case managers will plan services and support for individual elderly, and provide support to care-givers of the elderly (*Straits Times*, 13 January 2001).

Policy making decisions regarding health care for the elderly need to take into account changes in individual health status over time, and the prevalence of particular diseases. There are differences in how various cohorts of elderly report, and interpret, their health status. The presence of such differences highlights the importance of longitudinal studies that track individual changes in health over time.

Table 11 shows the percentage distribution of self-reported health status among elderly Singaporeans in 1995 and 1999. In 1995, 85 percent of older persons above 55 reported themselves to be in good health whereas less than one percent reported poor health status. In 1999, only 50 percent of older persons age 59 and above reported themselves to be in good health, and the percentage reporting poor health status was 13 percent. In both years, 1995 and 1999, the question asked regarding health status was identical, as were the answer categories provided. The elderly respondents were asked:

Q: How would you rate the state of your health at present?"
A: Very good
Good
Not too good
Poor

Source: Transitions in Health, Wealth, and Welfare of Elderly Singaporeans: 1995–1999.

There appears to be a large decline in self-rated health status between 1995 and 1999. These results are repeated in analysis of the longitudinal data available for the same time period.

Table 11. Health status of elderly Singaporeans, 1995 and 1999.

Health Status	1995 (55+)	1999 (59+)
Good	85.2	50.4
Not too good	14.0	37.0
Poor	0.8	12.6
Total (%)	100.0	100.0
N	4750	1981

Source: Transitions in Health, Wealth, and Welfare of Elderly Singaporeans: 1995–1999.

Looking at the same individuals over time, for one-third of elderly respondents, health remained good (see Table 12). However, for over half the respondents (53 percent) health status deteriorated. A small percentage of elderly, 9 percent, reported an improvement in health status. These results show that the health status of older persons age 55 in 1995 declined dramatically over the four-year period.

Research on the determinants of declines in health status reveals that being female elderly, Malay and Indian elderly, elderly with no formal education, those with lower monthly incomes, and elderly with little social support, are most likely to experience a decline in self-reported health status over time (Straughan, Chan and Teo 2001).

In general, the most common ailment among the current cohort of Singaporean elderly is high blood pressure or hypertension. However, the prevalence of certain diseases appears to be correlated with age and ethnicity. Among the young-old (59–69), Indians are most likely to suffer from diabetes compared to Chinese and Malays. Malays report higher rates of arthritis and chronic lung disease, whereas Chinese report higher rates of high blood pressure compared to the two other ethnic groups. Among the oldest-old (70 and above), Malay and Indian elderly report diabetes more frequently. Oldest-old Chinese are more likely to report high blood pressure and cancer compared to Malays and Indians (see Tables 13 though 15).

Table 12. Change in health status of respondents between 1995 and 1999.

Change in Health Status	Percentage
Health remained good	32.3
Health improved	8.5
Health declined	52.8
Health remained poor	6.4
Total	100

Source: Transitions in Health, Wealth, and Welfare of Elderly Singaporeans: 1995–1999.

Table 13. Type of ailments by age group for Chinese elderly.

Ethnic group	Ailments of respondents	Age categories for 1999 59–69	70–79	80+	Total
CHINESE	A Stroke	5.3%	10.0%	7.4%	7.3%
	High blood pressure/hypertension	47.0%	41.3%	38.0%	42.2%
	Diabetes	9.1%	7.0%	7.8%	8.1%
	Cancer or a maglignant	1.7%	1.2%	2.1%	1.7%
	Chronic lung disease	2.5%	4.0%	4.7%	3.7%
	Heart ailments	3.8%	4.6%	7.0%	5.2%
	Arthritis or rheumatism	25.6%	25.2%	23.0%	24.5%
	Permant loss of memory	.6%	.9%	2.3%	1.3%
	Kidney problems	.2%	.6%	.4%	.4%
	Cataract or glaucoma	4.0%	5.2%	7.4%	5.6%
	Total	472 100.0%	329 100.0%	487 100.0%	5.3% 47.0%

Source: Transitions in Health, Wealth, and Welfare of Elderly Singaporeans: 1995–1999.

These ethnic differences need to be examined more closely in future research. Cultural factors may play a part in the interpretation of disease and in determining the utilization of health services. The experience of disease may differ across ethnic groups and influence health-seeking behaviours.

Table 14. Type of ailments by age group for Malay elderly.

Ethnic group	Ailments of respondents	Age categories for 1999 59–69	70–79	80+	Total
MALAY	A stroke	5.6%		11.4%	5.6%
	High blood pressure/hypertension	42.3%	50.0%	17.1%	38.2%
	Diabetes	5.6%	15.8%	5.7%	8.3%
	Cancer or a maglignant		2.6%		.7%
	Chronic lung disease	8.5%		5.7%	5.6%
	Heart ailments	4.2%	10.5%	17.1%	9.0%
	Arthritis or rheumatism	29.6%	13.2%	22.9%	23.6%
	Permant loss of memory			2.9%	.7%
	Cataract or glaucoma	4.2%	7.9%	17.1%	8.3%
	Total	71 100.0%	38 100.0%	35 100.0%	144 100.0%

Source: Transitions in Health, Wealth, and Welfare of Elderly Singaporeans: 1995–1999.

Table 15. Type of ailments by age group for Indian elderly.

		Age categories for 1999			
Ethnic group	**Ailments of respondents**	**59–69**	**70–79**	**80+**	**Total**
INDIAN	A stroke	7.4%		20.0%	5.7%
	High blood pressure/hypertension	40.7%	47.6%		39.6%
	Diabetes	18.5%	19.0%		17.0%
	Cancer or a maglignant		4.8%		1.9%
	Chronic lung disease	3.7%			1.9%
	Heart ailments	11.1%			5.7%
	Arthritis or rheumatism	18.5%	23.8%	60.0%	24.5%
	Permant loss of memory			20.0%	1.9%
	Cataract or glaucoma		4.8%		1.9%
	Total	27 100.0%	21 100.0%	5 100.0%	53 100.0%

Source: Transitions in Health, Wealth, and Welfare of Elderly Singaporeans: 1995–1999.

Conclusion

The dramatic shift in age structure that Singapore will experience over the next 30 years was set in motion in the 1960s by an equally dramatic decline in fertility levels. These changes in the age structure are affecting economic and social aspects of life for all Singaporeans, among which the aspects of financial security, employment, living arrangements, and health care. This paper discusses the issues and policy implications for the family and state, in combination with recently available data on the elderly for Singapore.

The family in Singapore is under increasing pressure to support its older members. As a result of below replacement fertility rates since 1977, the number of children available to support older parents has declined dramatically. The current generation of elderly in Singapore have an average of four children to support them. The next generation of elderly will have half that number or fewer. In the case of limited numbers of family members to support older parents, what role should state policy play? The Singapore government continues to see the family as the main provider of support for the elderly. At the same time, policies for the aged are being revised to take into account changes in elderly characteristics over time. For example, future elderly will be more educated and have higher incomes. Therefore, CPF investment opportunities are being expanded to allow for more individual decision-making. The raising of the retirement age is another indication that the government is interested in tapping the potential of elderly workers, and at the same time, promoting individual responsibility for financial support in old age.

In terms of financial security, the current generation of elderly have minimal amounts (less than $5, 000) in their CPF accounts. Over time, the amount of actual income received by elderly varies significantly, as does the perception of income adequacy. This points to the need for policies that are sensitive to the changing financial needs of the elderly over time rather than viewing these needs as static. The longitudinal data also reveal changes in living arrangements over time although the changes are not as dramatic as those for income. A significant amount of this change in living arrangements is the result of adult children leaving the home.

With regards to employment among this current generation of elderly, the majority retired during the two waves of the survey. We do see, however, an increasing percentage of elderly persons who are not working but looking for work. This suggests that policies aimed at hiring older workers will be well received by the elderly themselves in future. The success of these policies, however, requires a change in mindset among many employers who retain ageist attitudes.

Finally, analyses of self-reported health status among this elderly cohort reveals large declines in health status that occurred between 1995 and 1999. Declines in health status are significantly correlated with a lack of formal education, lower income, and a lack of social support. These findings suggest sub-groups of elderly that require special attention with regard to health care.

In conclusion, findings from research using these longitudinal data reveal a substantial amount of change in the elderly individual's economic status, employment status, living arrangements, and health status, over time. These results point to the need to develop policies that cater to specific sub-groups of elderly persons. The implementation of uniform policies across all ages of elderly runs the risk of neglecting needs of specific sub-groups over time. The formulation of age-specific policies requires an in-depth understanding of the processes that occur during old age. This includes understanding how individual processes interact with the family and society. This can be achieved through the analysis of longitudinal data that provides information on individual changes over time.

References

Andrews, G.R. and Hermalin A.I. (2000), "Research Directions in Ageing in the Asia-Pacific Region: Past, Present and Future." In: *Ageing in the Asia-Pacific Region: Issues, Policies and Future Trends*, D.R. Phillips, ed. Routledge: London.

Asher, M. (1996), "Financing old age in Southeast Asia: an overview." *Southeast Asian Affairs, pp 70–98*, Central Provident Fund Board website at www.cpf.gov.sg.

Chan, A. (1997), "An overview of the living arrangements and social support exchanges of older Singaporeans." *Asia-Pacific Population Journal*, 12(4):35–50.

Chan, A. (1999), "The social and economic consequences of ageing in Asia." *Southeast Asian Journal of Social Science*, 27(2):1–8.

Chan, A., Ofstedal, M.B. and Hermalin, A.I. (2002), "Changes in subjective and objective measures of economic well-being and their interrelationship among the elderly in Singapore and Taiwan." *Social Indicators Research*, 57:263–300.

Chang, M-C. (1999), "A longitudinal study on living arrangements of the elderly in Taiwan." In: *Emerging Social Economic Welfare Programs for Ageing in Taiwan in a World Context*, C. Chen, A.I. Hermalin, S-C. Hu and J.P. Smith, eds. Institute of Economics, Academia Sinica: Taipei.

Choon, A.T. and Low, L. (1996), "Social security: How Singapore does it?" *Asia-Pacific Journal of Social Work*, 6(1):97–119.

Cowgill, D.O. and Holmes, L.D., eds. (1972), *Ageing and Modernization*, Appleton-Century-Crofts: New York.

Fawcett, J.T. and. Khoo, S.E. (1980). "Singapore: rapid fertility transition in a compact society." *Population and Development Review*, 6(4):549–579.

Frankenberg, B., Chan, A. and Ofstedal, M.B. (2002), "Stability and change in living arrangements of the elderly in Southeast Asia." *Population Studies*, 26:201–213.

Hermalin, A. (1997), "Drawing policy lessons for Asia from research on ageing." *Asia Pacific Population Journal*, 12(4):89–102.

Hermalin, A. (2000), "Challenges to comparative research on intergenerational transfers." *Comparative Study of the Elderly in Asia Research Report*, No. 00–56. Population Studies Center, University of Michigan.

Inter-Ministerial Committee on Ageing Report. (1999), Ministry of Community Development: Singapore.

Jones, G.W. (2000), "Human capital aspects of economic development: a comparative perspective in Asia." Paper presented at the *IUSSP Conference on Age Structural Transitions and Policy Implications, Nov. 8–10, Phuket, Thailand.*

Kinsella, K. (2000), "Demographic dimensions of ageing in East and Southeast Asia." In: *Ageing in the Asia-Pacific Region: Issues, Policies and Future Trends*, D.R. Phillips, ed. Routledge: London.

Knodel, J. and Nibhon, D. (1997), "Living arrangements and support among the elderly in South-East Asia: An Introduction." *Asia Pacific Population Journal*, 12(4):5–16.

Knodel, J. and Nibhon, D. and Chayovan, N. (1997), "Family support and living arrangements of Thai elderly." *Asia Pacific Population Journal*, 12(4):51–68.

Lee, S.M., Alvarez, G. and Palen, J.J. (1991), "Fertility decline and pronatalist policy in Singapore." *International Family Planning Perspectives*, 17(2):65–73.

Leete, R. (1994), "The continuing flight from marriage and parenthood among the overseas Chinese in East and Southeast Asia: dimensions and implications." *Population and Development Review*, 20(4):811–829.

Martin, L. and Kinsella, K. (1994), "Research on the demography of ageing in developing Countries." In: *Demography of Ageing*, L. Martin and S.H. Preston, eds. National Academy Press: Washington, DC.

Natividad, J.N. and Cruz, G. (1997), "Patterns in living arrangements and familial support for the elderly in the Philippines." *Asia Pacific Population Journal*, 12(4):17–34.

Ofstedal, M.B., Chayovan, N., Chan, A., Chuang, Y.L., Perez, A. and Hermalin, A.I. (2001), "Policies and programs pertaining to the welfare of the elderly." In: *The Well-Being of the Elderly in Asia: A Four-Country Comparative Study*, A.I. Hermalin, ed. University of Michigan Press: Ann Arbor, MI.

Ofstedal, M.B., Chayovan, N., Chan, A., Chuang, Y.L., Perez, A., Hermalin, A.I., Knodel, J. and Chayovan, N. (1999), "Intergenerational support and gender: a comparison of four Asian countries." *Southeast Asian Journal of Social Science*, 27(2):21–42.

Phillips, D.R. (2000), "Ageing in the Asia-Pacific region: issues, policies and contexts." In: *Ageing in the Asia-Pacific Region: Issues, Policies and Future Trends*, D.R. Phillips, ed. Routledge: London.

Pool, I. (2000), "Age-structural transitions and policy: frameworks." Paper presented at the *IUSSP Conference on Age Structural Transitions and Policy Implications, Nov. 8–10, Phuket, Thailand.*

Rappa, A. (1999), "The politics of ageing: perspectives from state and society in Singapore." *Southeast Asian Journal of Social Science*, 27(2):123–138.

Saw, S.H. (1975), "Singapore: resumption of rapid fertility decline in 1973." *Studies in Family Planning*, 6(6):166–169.

Saw, S.H. (1999), *The Population of Singapore*, Institute of Southeast Asian Studies: Singapore.

Shantakumar, G. (1994), *The Aged Population in Singapore*, Singapore National Printers: Singapore.

Shantakumar, G. (1999), "Ageing in the city-state context: perspectives from Singapore." *Ageing International, Summer*, pp 46–60.

Straits Times. 2000, *More Babies Wanted: Bonus for Second and Third*, Singapore Press Holdings: Singapore.

Straughan, P., Chan, A. and Teo, P. (2001), *The Determinants of Changes in Perceived Health Among Older Singaporeans*, National University of Singapore: Singapore. (Working draft.)

Vasso, S., Ngiam, T.L. and Cheung, P.L. (2000), "Singapore's ageing population: social challenges and responses." In: *Ageing in the Asia-Pacific Region: Issues, Policies and Future Trends*, D.R. Phillips, ed. Routledge: London.

World Bank. (1994), *Averting the Old Age Crisis: Policies to Protect the Old and Promote Growth*, Oxford University Press: Oxford.

INDEX